Cyber-Physical Energy and Power Systems

Yijia Cao · Yong Li · Xuan Liu ·
Christian Rehtanz

Cyber-Physical Energy and Power Systems

Modeling, Analysis and Application

Springer

Yijia Cao
College of Electrical
and Information Engineering
Hunan University
Changsha, Hunan, China

Yong Li
College of Electrical
and Information Engineering
Hunan University
Changsha, Hunan, China

Xuan Liu
College of Electrical
and Information Engineering
Hunan University
Changsha, Hunan, China

Christian Rehtanz
LS Energiesysteme und Energiewirtschaft
TU Dortmund University
Dortmund, Nordrhein-Westfalen, Germany

ISBN 978-981-15-0061-9 ISBN 978-981-15-0062-6 (eBook)
https://doi.org/10.1007/978-981-15-0062-6

This Springer imprint is published by the registered company Springer Nature Singapore Pte Ltd.
The registered company address is: 152 Beach Road, #21-01/04 Gateway East, Singapore 189721, Singapore

Preface

With the advancement of smart grid and ubiquitous power Internet of things strategy, much electrical equipment, data acquisition equipment, and computing equipment are interconnected via the grid and communication network. The traditional energy and power system with physical equipment as the core has gradually evolved into a highly coupled cyber-physical energy and power system (CPEPS). Generally, CPEPS is a new type of system in which the traditional energy system is integrated into the information network with control, communication, and computation functions.

CPEPS is the basis of the long-term transformation of energy management, which will profoundly change the perspective and application pattern of traditional energy research. However, considering that the physical system and the information system are separated from each other in the traditional power system modeling, analysis, optimization, and control method, this cannot meet the requirements of smart grid development. In this book, from the aspects of system modeling, analysis, and application, we build the integrated framework that reflects the characteristics of energy and information systems, explore the integration mechanism, and realize the unification of energy system and information system.

This book intends to report the new results of modeling, analysis, and application in CPEPS. It collects new research ideas and achievements such as a cascading failure model for CPEPS, a quantitative analysis method of data flow for a typical substation, a reliability analysis method for substations with a cyber-physical interface matrix, a simplified co-simulation model for analyzing the interdependencies between the CPEPS, an architecture of co-simulation environment based on JADE framework, and an optimal attack strategy in CPEPS.

The first motivation of this book is to establish a systematic, efficient, and comprehensive approach to analyze the interaction impact and build the co-simulation framework for CPEPS with multi-source energy flow and heterogeneous information flow. A cascading failure model is proposed with one-to-multiple interdependency, and a relevant theoretical framework is proposed to analyze CPEPS, which can reveal the coupling and interaction mechanisms of information flow and energy flow. Furthermore, it is significant and mature to build the co-simulation framework and obtain the quantitative relation between information features and energy system operations.

The second motivation of this book is to carry out the systemized research on reliability analysis and cyber-attack security for CPEPS. A reliability analysis method for substations is sufficiently developed with a cyber-physical interface matrix, which aims to calculate the influences from both the physical device failures and the communication device failures. To study the growing number of cyber-attack influence on CPEPS, it is of vital importance to investigate the cyber-attack on energy systems and explore the optimal attack strategy from the attacker's perspective.

The main research results of this book are original from authors who carry out the related research together, which is a comprehensive summary of authors' latest research results. This book is likely to be of interest to university researchers, R&D engineers, and graduate students in electrical engineering who wish to learn the core principles, methods, algorithms, and applications of CPEPS.

Outlines

This book is divided into 11 chapters. Chapter 1 introduces the status quo and trends of the fusion of cyber and power systems, the critical scientific problems and technologies in the field of CPEPS, and the interaction mechanism and modeling methods of CPEPS. Moreover, the mass data processing and cluster analysis, the architecture of a communication network, the information transmission technology, and security of CPEPS are also summarized and analyzed.

Chapter 2 analyzes the approximation to describe the interdependence in CPEPS based on the dynamic power flow model. The topological and partial transmission characteristics of dispatching data networks are considered. By investigating the structural imparts of dispatching data networks on load shedding in case of different attacks on power grids, in most cases, the double star structure is better than the meshed one.

Chapter 3 introduces a model of cyber-physical systems in cascading failure situations. By introducing the control threshold, the cascading failure model with the existence of redundancy and standby lines in control supply is developed. With the one-to-multiple interdependent relationship, the critical point based on percolation theory is measured to evaluate the robustness of CPEPS. For communication network, the Small Word (SW) network and the Double Star (DS) network are compared to find the robustness impact on topology, interdependent link, and control threshold.

Chapter 4 summarizes the principles of two typical protection algorithms for distribution networks, designs the calculation process of differential protection based on the Ethernet passive optical network (EPON), and analyzes the impacts of EPON-based communication networks on the differential protection. This designed method can protect the distribution network effectively. In the system with time synchronization, the current differential protection (CDP) can achieve precise fault location and isolation, while the directional comparison pilot protection (DCPP) has better reliability in the system without time synchronization.

Chapter 5 proposes three kinds of mathematical models as typical data flow within substations according to IEC 61850, which are cyclic data, stochastic data, and burst data. Thereby, a quantitative analysis of data flow is carried out for a typical substation based on the proposed data models. The advantage of VLAN and impacts of system faults as well as network topologies on a VLAN-based network are also evaluated and simulated by OPNET Modeler.

Chapter 6 proposes a reliability analysis method for substations with a cyber-physical interface matrix (CPIM). This strategy calculates the influences from both the physical device failures and the communication device failures. The simplified model of the practical substation based on the Chinese IEC 61850 standard is used.

Chapter 7 presents a new vision for the distribution system where prosumers are encouraged by different balancing premiums in a local community. Price-responsive generation and individual demand are affected by his/her attitudes and inherent characteristics. A load aggregator participates in the community market that runs by a local electricity coordinator. A regulator is assumed to design balancing premium schemes. And a multi-agent-based simulation with a four-layered representation is employed to study features of the community and incentive strategies for the desired performance.

Chapter 8 categorizes the potential impacts on the information flow respecting the end-to-end information features in terms of delayed, disordered, dropped, and distorted data. Then, a simplified co-simulation model is introduced for analyzing the interdependencies between energy and information flows, and obtaining the quantitative relation between information features and power system operations. This co-simulation model features in low complexity and covering potential cyber contingencies. Moreover, the quantitative relation obtained by the model can provide requirements for the planning and operation of CPEPS.

Chapter 9 proposes a JADE-based information physical system co-simulation environment to analyze and simulate physical state of power flow and transient fault of power grid for smart distribution networks. The characteristics of the simulation environment are discussed, and the modeling methods of distributed controllers are described in detail. This environment can realize the connection of the software PowerFactory and OMNeT++, and is suitable for the co-simulation of distributed system and MAS. By analysis of the results of the controllers, the effectiveness and deficiency of algorithm can be evaluated easily, and the effectiveness of the proposed co-simulation environment is validated.

Chapter 10 proposes a local attack model based on incomplete network information. In this model, we show that the attackers only need to obtain the network information of the local attacking region to inject false data into smart meters in the local region of the power grid without being detected by the state estimator. In addition, four attacking cases are analyzed in detail. Simulations on the modified IEEE 14-bus system confirm the effectiveness of the proposed model and algorithms.

Chapter 11 analyzes the topological characteristics of an attacking region and builds the corresponding mixed integer linear programming (MILP) model. Based on the local load redistribution (LR) attack theory, the optimal attack region of single load bus is determined by obtaining reduced network information. According to the topological characteristics of the attack region, the optimal attack region is determined by the detailed steps. Based on the attack principle, heuristic algorithm is adopted to determine the optimal attack area load bus.

Changsha, China Yijia Cao
Changsha, China Yong Li
Changsha, China Xuan Liu
Dortmund, Germany Christian Rehtanz

Acknowledgements

This book project is supported by the National Natural Science Foundation of China (NSFC) under Grant 51520105011 and 51822702, by the Key Research and Development Program of Hunan Province of China under Grant 2018GK2031, by the 111 Project of China under Grant B17016, and by the Excellent Innovation Youth Program of Changsha of China under Grant KQ1802029.

The authors would like to thank Prof. Xiaoxin Zhou (China Electric Power Research Institute), Prof. Shijie Cheng (Huazhong University of Science and Technology), Prof. Tapan K. Saha and Dr. Olav Krause (University of Queensland), Prof. Ryuichi Yokoyama (Waseda University), Prof. Kwang Y. Lee (Baylor University), Prof. Mohammad Shahidehpour (Illinois Institute of Technology), Dr. Ulf Häger (TU Dortmund), and Dr. Yi Tan (Hunan University) for their great support and valuable comments.

Special thanks to the postgraduate students Mr. Junjie Zhong, Mr. Yirui Zhao, Ms. Yahui Wang, Mr. Bonan An, Mr. Shaoyang Wang, Mr. Li Jiang, and Mr. Guofeng Hu for their contribution and proofreading. Finally, each author would like to thank the long-term support and encouragement from their family.

Changsha, China — Yijia Cao
Changsha, China — Yong Li
Changsha, China — Xuan Liu
Dortmund, Germany — Christian Rehtanz
2019

Contents

Chapter 1
Introduction

Cyber-physical energy and power system (CPEPS) combines the computation, communication, and control technologies with physical power systems and realizes the efficient fusion of power, information, and control [1]. In this chapter, we summarize and analyze the related critical scientific problems and technologies by promoting the development of CPEPS. Firstly, the co-simulation platform of CPEPS and its evaluation is overviewed because the co-simulation is an effective method to investigate infrastructure interdependencies. Then, this chapter gives the analysis of the critical problem of CPEPS, namely the interaction between energy and information flows, especially the influence of the information communication technology (ICT) failures on power systems. Also, the related analysis methods are summarized. After that, different control principles and the concept of distributed coordination control in the information network environment are outlined and illustrated. Besides, mass data processing and cluster analysis, the architecture of the communication network, information transmission technology, and security of CPEPS are summarized and analyzed.

1.1 Status Quo and Trends of the Fusion of Cyber and Power Systems

Because the smart grid has the potential to utilize sustainable energies sufficiently and to improve the safety, reliability, and efficiency of the power grid, its development has attracted more and more attention. Information technology plays a crucial role in smart grids [2, 3], and the high-speed and bidirectional communication infrastructure, advanced information processing, and distributed computing technologies are indispensable for system state estimation, control, optimization, and self-healing of smart grids. The power system operation depends on not only energy flow but also information flow. The problems of information safety, mass data processing, communication reliability, and so on will generate profound influence on grid operation.

Y. Cao et al., *Cyber-Physical Energy and Power Systems*,
https://doi.org/10.1007/978-981-15-0062-6_1

In the literatures [4–7], intensive studies regarding traditional power grid modeling, analysis, and control methods were carried out, which can be used to study the influence and correlations between information system and energy power system.

CPEPS is a power and information integrated system. Information communication technologies play an essential role in CPEPS. The devices of sensor, computation, and control units are connected by the communication network to realize the information sharing in whole systems, and together with distributed computation technology, the identification, optimization, and control of the physical system can be performed. The safety, reliability, and efficiency of the power system can *be improved* by the above fusion of power, information, and control, which provides physical entities with functions of computation, communication, accurate control, coordination, and autonomy [8]. Moreover, CPEPS can cooperate with other social systems, like the transportation system, and with the environment to realize the green economy and sustainable development.

As shown in Fig. 1.1, CPEPS can monitor and control power system in a secure, reliable, and efficient way, which is helpful to achieve the optimal balance between

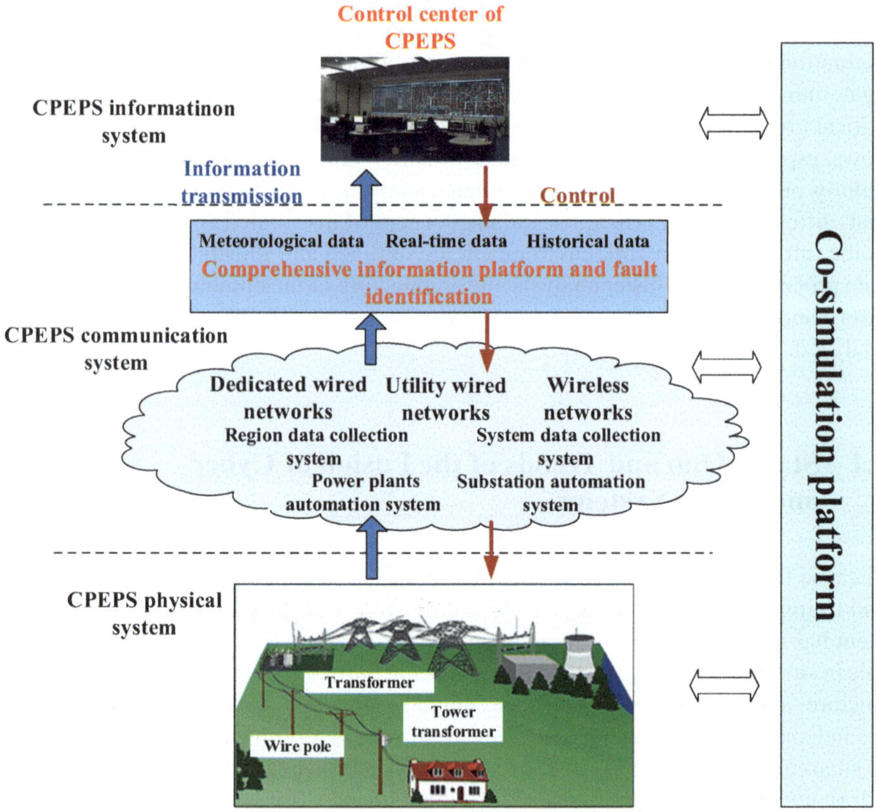

Fig. 1.1 Configuration of multi-objective robust controller synthesis

the generation, distribution, and consumers [9]. However, its development still faces many problems, such as the co-simulation platform establishment and evaluation, the interaction between power and information flow, and the power system control and protection.

1.2 Simulation and Evaluation Methods and Its Application in CPEPS

Simulation is a common method to evaluate the performance of CPEPS. Currently, both the power and the communication systems are analyzed by using separate simulators. For simulating the whole CPEPS, there are two ways that one is re-implementing the communication models in power system simulator (or vice versa), and another is establishing a co-simulation platform which combines both simulators and using simulation control module to realize time synchronization and data exchange.

Regarding the re-implementation method, the time synchronization and data reconstruction are not needed, but power system simulator usually cannot support exhaustive functions needed by the communication system modeling or vice versa. In [10], an integrated development platform was set up in communication simulator OMNeT++. The power system was modeled in MATLAB and then linked to OMNeT++. The platform accomplished static power system analysis. Due to the limitation of simulators, it is difficult for dynamic power system simulation. Co-simulation is able to ensure the simulation accuracy and efficiency by directly using dedicated and available libraries for both power and information communication systems [11].

The co-simulation mainly consists of the following four parts: (1) power system simulation, (2) information system simulation, (3) communication network simulation, and (4) simulation control module, as illustrated in Fig. 1.2. The power system simulation is developed to account for relevant components interacting in the operation of the power system. The information system simulation mainly contains two functions, that is to say, data analysis and computation, and system control and protection. The imitation of a practical communication network, which exchanges data between the power system and the information system, is completed by communication network simulation. The simulation control module is responsible for the coupling between different simulators by defining data exchange and time synchronization mechanisms.

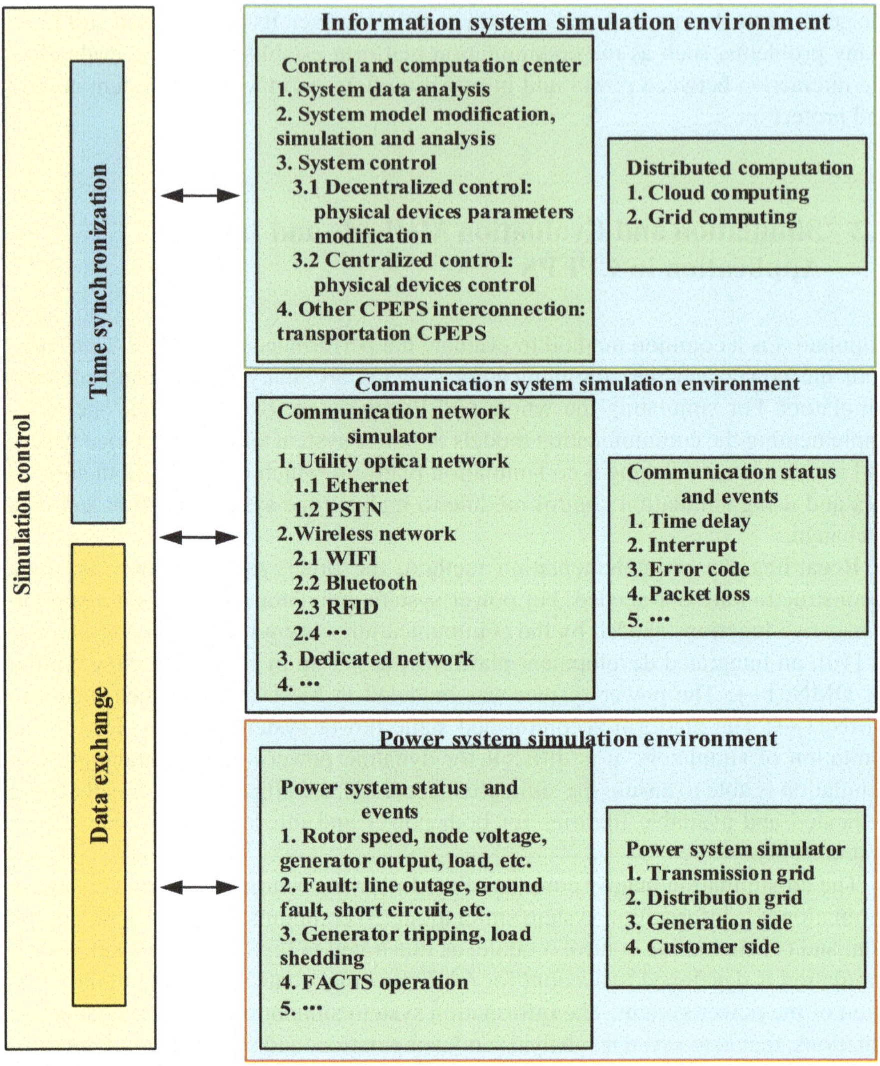

Fig. 1.2 Co-simulation architecture for CPEPS

1.2.1 Power, Communication, and Information System Simulation

Commonly used simulation tools for the power system simulation are DIgSILENT/PowerFactory [12], PSCAD/EMTDC [13], MATLAB [14], Adevs [15], Modelica [16], PSLF [17], OpenDSS [18–20], VTB [21–23]. In the literature [12], for real-time evaluation of cyber-physical energy system, a co-simulation environment called INSPIRE has been established, and DIgSILENT/PowerFactory is employed for simulating the electromechanical dynamics of the power system. In the literature [13], for evaluating the outcome of the proposed communication strategy on fault monitoring of the power system, the IEEE 13-node test feeder model was set up within PSCAD/EMTDC. To evaluate the performance of wide-area damping control of the power system, an IEEE benchmark test system was modeled in MATLAB [14].

By modeling network topology, communication protocol, and so on, in communication network simulators, the communication system simulation realizes the data exchange between power and information systems. Besides, mathematical models for the data flow [24], the estimation for the delayed or lost data [25], the data authentication [26], and so on are also important to implement the information communication between cyber and power systems. Commonly used communication network simulators are NS-2 [15–17, 27], OPNET [20–23, 28, 29], and OMNET++ [10, 19].

The information system works in either decentralized or centralized level, and implements data analysis, model identification, algorithm generation of system control and protection, as well as distributed computation [30, 31]. Commonly used tools to model the information system are MATLAB/ Simulink [32], Microsoft Visual Studio [13], and Java-based agents [33].

1.2.2 Simulation Control

Normally, power system simulator is time-driven, while communication system simulator is event-driven. Hence, it is necessary to design time synchronization and data exchange for the co-simulation. At present, there are three synchronization methods, i.e., master–slave [10, 16, 19, 29], time-stepped [21–27], and global event-driven methods [15, 18], as shown in Fig. 1.3 [34]. In the master–slave method, master simulator coordinates the co-simulation. During co-simulation, the master simulator has higher priority and sends time synchronization comment to slave simulator, when data exchange is needed. Either the power system simulator or communication simulator can be regarded as the master. Communication system simulator is the master in the figure. Regarding the time step synchronization method, simulators run their simulation independently and only pause at the defined synchronization time to exchange data. Regarding the event-driven synchronization method, a global event list is prepared, and according to their time stamp, simulators run orderly.

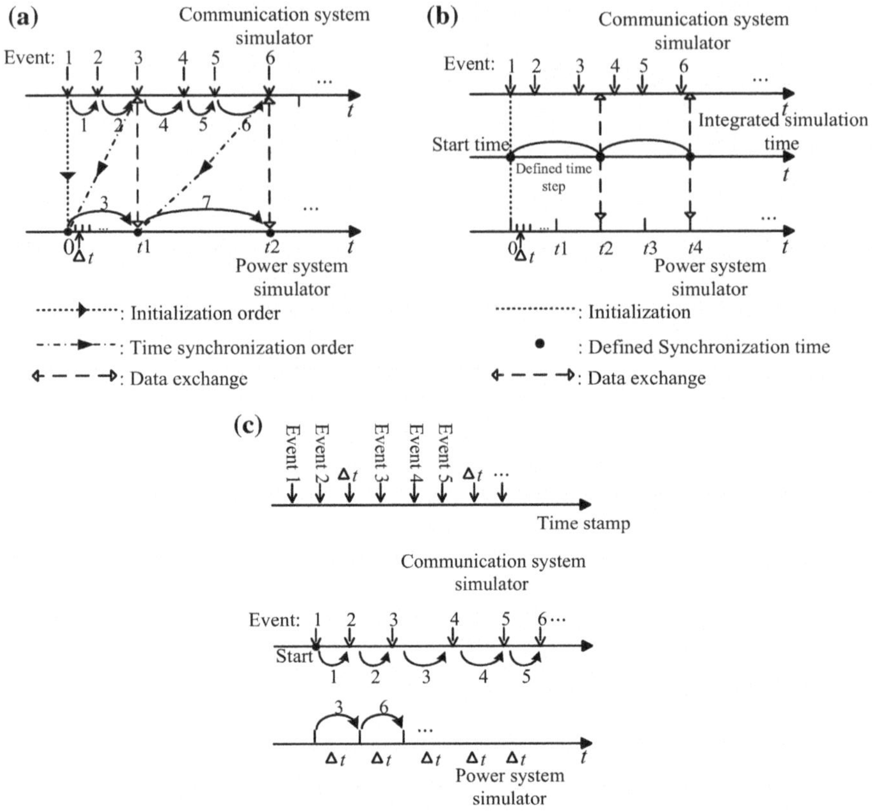

Fig. 1.3 Synchronization methods. **a** Master–slave synchronization method. **b** Time step synchronization method. **c** Event-driven synchronization method

1.3 Interaction of CPEPS and Related Analysis Methods

1.3.1 Interaction Between Energy and Information Flows

The dynamic process of power system control is speeded up for the more and more renewable power generations and flexible load, which makes the assistance from ICT indispensable [35, 36]. The introduction of ICT renders the CPEPS an integrated system which combines the power and information and communication systems. The interaction among these systems has to be considered carefully for the CPEPS.

Questions have been raised about the interaction between the power system and ICT when faults happen in any of them, as illustrated in Fig. 1.4. In [35], it summarized some of the most important ICT failures and their effects. When power system faults happen, there is huge data needed to be collected by the control center, including node voltages, power flow on the transmission line, and flexible load

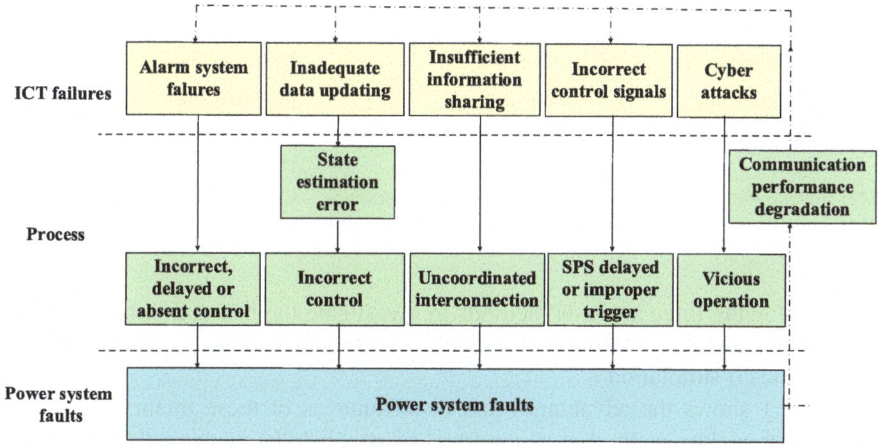

Fig. 1.4 Interaction between power system and ICT

states. The boomed communication traffic will degrade the performance of communication network and induce the increased packet loss and time delay as well as decreased transmission rates, which may lead to ICT failures reversely and aggravate the impact of power system faults. The literatures [36, 37] performed some works in investigating the interdependencies between energy and information flows. In [36], an interactive cascading model is proposed for power grids and coupled communication systems, which is based on the redistribution of DC power flow and the routing of the Open Shortest Path First strategy. By the cascading model and related quantification methods, it was found that when the power system is at its self-organized criticality, a catastrophic cascading failure in the power system can be indicated by the threshold value of the communication network transmission inefficiency. In [37], the interaction between energy and information flows is modeled by a dynamic power flow, and it was found that under random attacks, the double-star dispatching data network has a lower probability of catastrophic failures, and under intentional attacks, the mesh network has advantages in transmission performance.

1.3.2 Analysis Methods

In CPEPS, the power system is a continuous dynamic system, and the ICT is a discrete event-based system. For the continuous system, its analysis and modeling methods are based on continuous mathematical theory, and differential-algebraic equations are often used for system modeling. However, for the discrete system, the discrete mathematical theory is its basis, and the common modeling tool is finite automation. For the continuous–discrete coupled system, the theoretical basis, modeling tools, and uncertain interaction are major challenges for system modeling. At present, there

Table 1.1 Comparison of different analysis methods

Analysis methods	Advantages	Disadvantages
Dynamical modeling	Theoretical study	Model mismatching problem
Coupled network modeling	More details on interaction	Time-consuming modeling process
Real-time co-simulation	More details on evolutionary process of system vulnerability	Cross-discipline software utilization

are the following three analysis methods to investigate the interaction between the power system and ICT: (1) dynamical modeling, (2) coupled network modeling, and (3) real-time co-simulation.

Table 1.1 shows the advantages and disadvantages of these methods. For the dynamical modeling method, it is important to establish the mathematical formulations of both the information and the power flow. Based on this, the system characteristics, such as stability, controllability, and observability, can be investigated. In [38], a unified modeling method was proposed based on mathematical tools such as differential-algebraic equations, finite automation, stochastic process, queueing theory, and so on. Moreover, the information flow model considers various ICT devices such as the router, communication line, and sensor units.

For the coupled network modeling, it firstly analyzed the statistical properties of power and information systems. These properties include failures in time rates, mean time between failure values and transition time, and the finite time necessary for events to occur. Because it is necessary to clarify static statistical properties and the coupled relation of each device in power and information systems, the time-consuming modeling process is the main problem. In [39], a method using marked Petri net models was proposed to simultaneously model the power and the telecommunication infrastructures. The established model can be used to analyze how events in the communication infrastructure interact with the power infrastructure and result in power blackouts.

The real-time co-simulation can be used to analyze the evolutionary process of the system vulnerability. By deriving the mathematical expression of various events emerging forms in power and information systems and quantitative index of vulnerability, the influence of emerging forms on system stability can be evaluated during the real-time simulation. But the ability of utilizing both power and information system simulators is the main challenge. Various co-simulation frameworks were proposed in [10, 15–29].

1.4 Challenges of Power System Control and Protection in CPEPS

Although the networked control can be used with the sufficient utilization of global system information to achieve stable and efficient operation of the power system, control devices have to face problems of uncertainty in time delay and information path, which impairs the controllability and observability of system and leads to instability and collapse of the whole system [40–42]. Networked and local combined control [43], and time delay and data loss compensation control [41, 42] are two main solutions.

For the networked and local combined control, the performance index is firstly proposed. After that, the control system is switched between networked and local modes according to the performance index to ensure the stability and reliability of the power system. Although the networked control can achieve better control performance, when the information system works normally, the low-performance information system will degrade the performance of the networked control, even worse than the local control which only needs the local information and reduces the demand on the communication network. In this case, the local control is activated. In [43], an allowable time delay named equivalent time delay (ETD) was proposed for the ICT infrastructure, as shown in Fig. 1.5, which means the performance of networked control with the ETD is equivalent to that of local control. Therefore, only when the total time delay in the networked control is less than the ETD, the networked control is used.

For the time delay and data loss compensation control, one most used principle is to adopt receding horizon optimization to calculate control signals of current and future several time intervals for controllable devices [38]. When actuators cannot receive the current control signal due to information network problems, they perform predicted control signals received before.

Since there are a large number of physical devices in CPEPS, the centralized control is impractical. The distributed coordination control, as shown in Fig. 1.6, provides

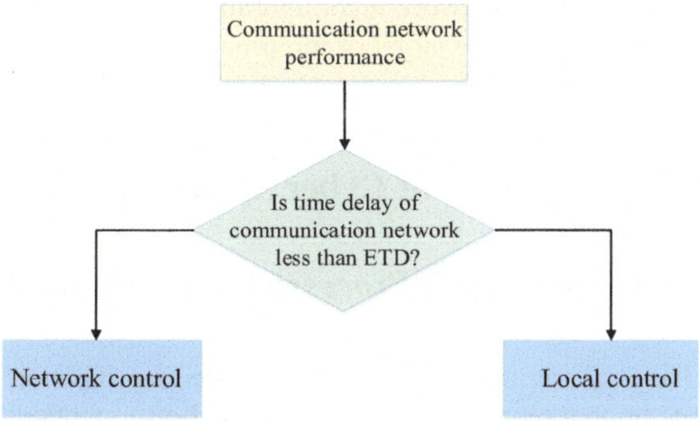

Fig. 1.5 Schematic diagram of networked and local combined control

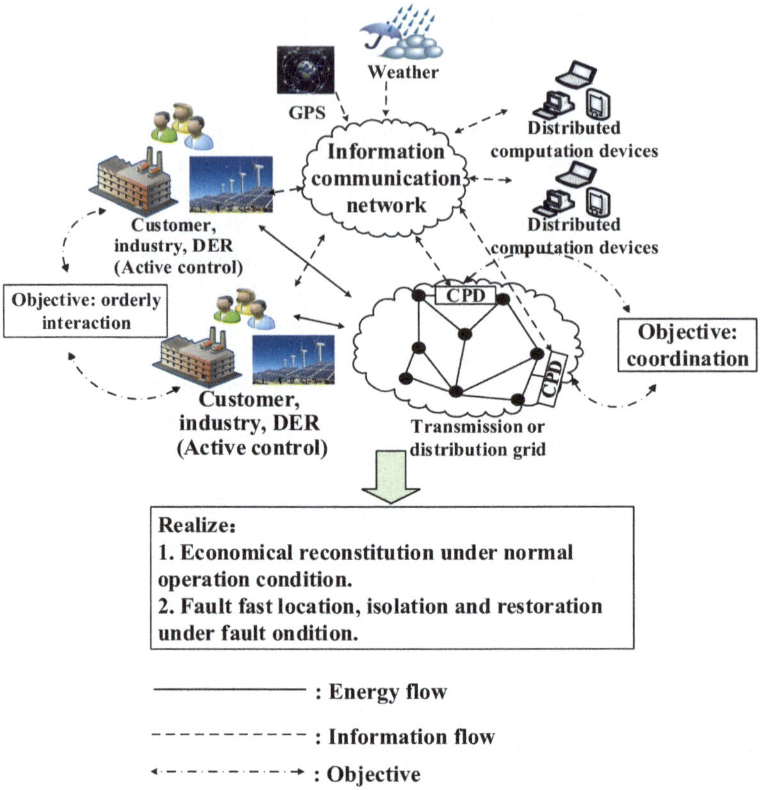

Fig. 1.6 Distributed coordination control concept in CPEPS

an efficient way. The bidirectional information is a media to trigger the orderly interacting and the active control of customer, industry, and distributed energy resources (DERs). The orderly interacting and the active control, together with the coordination between distributed control and protection devices (CPDs), can be used to realize the economical network reconfiguration as well as fast fault location, isolation, and restoration with the target of minimizing the area of a power blackout.

1.5 Challenges of Cyber Systems in CPEPS

1.5.1 Mass Data Processing and Cluster Analysis

Using mining mass data of CPEPS, the identification and assessment of the secure operating region, critical region of instability, and fault operation region of the power grid can be realized. Besides, the following issues also should be paid attention:

(1) fault information transmission mechanism in temporal and spatial multi-dimensional scale; (2) fault feature mining and interference signal identification theory; and (3) detection method of latent and high resistive faults, and fault accurate location theory.

1.5.2 Architecture of Communication Network

The architecture can be divided into three levels: plant/substation, region, and system levels. In each level, the star–mesh network, as shown in Fig. 1.7, is employed to consider the reliability and economy. The lowest level consists of power plants, substations, smart loads, and control and protection devices and takes the responsibility to collect real-time data and perform control and protection. The regional control center in the second level aims at the control and protection in its region and exchanging data with the system control center. The top level is the system control center, which collects, analyzes system data, and coordinates each control substation to realize the global optimization.

There are three types of communication networks, i.e., the dedicated wired, the utility wired, and the wireless networks [44]. The dedicated wired network has advantages of low time delay and high transmission reliability and can be utilized to connect the control center with critical sensors and control devices. The utility wired network should also be considered for considering the economy of a communication network.

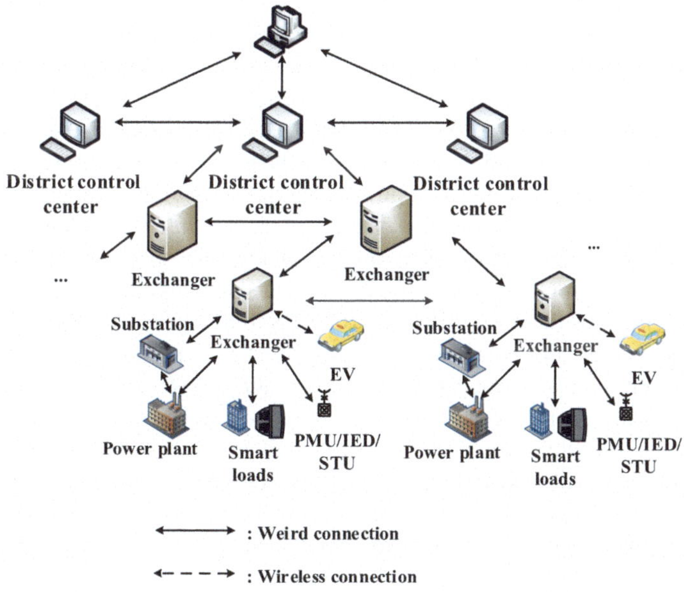

Fig. 1.7 Communication network architecture

However, for the problems of time delay, packet loss, and so on, it should be in charge of the communication of non-critical devices, like computation and backup devices. The wireless network is a feasible way for mobile components of the system, like electric vehicles (EVs) and devices where the communication wire is hard to access [45].

1.5.3 Information Transmission Technology

The communication traffic will inevitably increase for more and more involved devices in CPEPS. Due to the capacity limitation, the increasing communication traffic gives rise to network congestion and degrades the transmission performance of the network, which results in transmission speed decrease and time delay increase. Thus, it is necessary to investigate congestion and traffic control, and the equilibrium approach of real-time information flow. Information scheduling approach should be formed to address the soar in communication traffic when faults or other urgent cases happen in the grid. Besides, the information security, defense mechanism, communication protocol, network standard, and compatible information model for various devices are other fields should be focused on.

1.5.4 Security of CPEPS

Because CPEPS is a physical and information coupled system, the security of CPEPS should consider both physical and information systems. Besides, traditional security problems of the power system, such as transient stability, angle stability, security problems of information system in information communicating, processing, decision-making process, and so on, should also be paid close attention to. Moreover, due to the close interaction between physical and information systems, the security problems of power and information systems are closely interdependent [46]. Faults and cyber-attacks in information system will influence the security of power system or vice versa; for example, when the modern power system is attacked by a hacker, an attack which falsifies the data will render CPEPS to make a wrong control decision which damages the security of power system. Moreover, it will cause huge harm to the entire system such as out service of communication system and the mask of overloading lines and generators, and the whole system could become unstable in some extreme cases. The literatures [47–50] performed some works in power and information system security and risk assessment, which lays the foundation to analyze the correlation of the power security and information security, and can be utilized to establish the power and information combined security theory for CPEPS.

1.6 Summary

This chapter analyzes the critical scientific problems and technologies in the field of CPEPS. For the simulation and evaluation of CPEPS, co-simulation is an efficient way which directly uses dedicated and available libraries to ensure the simulation accuracy and efficiency. Various simulators of the power, the communication and the information systems, and the synchronization methods in simulation control are overviewed. Since CPEPS is a highly integrated system, the interaction between the power system and ICT, especially, the interaction when faults happen in any part of them, has to be considered carefully. Therefore, the interaction mechanism and modeling methods are analyzed and summarized. In the context of the information network, the stable and efficient operation of the power system can be achieved. However, the uncertainty in ICT should be considered in the power system control and protection. To address the uncertainty, two main solutions, i.e., networked and local combined control and time delay and data loss compensation control, are used. Moreover, the mass data processing and cluster analysis, the architecture of a communication network, the information transmission technology, and security of CPEPS are also summarized and analyzed in this chapter. By solving the above problems and advancing the technologies, the development of CPEPS will be significantly promoted.

References

1. Shi X, Li Y, Cao Y, Tan Y (2015) Cyber-physical electrical energy systems: challenges and issues. CSEE J Power Energy Syst 1(2):36–42
2. Ilic MD, Xie L, Khan UA, Moura JMF (2010) Modeling of future cyber–physical energy systems for distributed sensing and control. IEEE Trans Syst Man Cybern A Syst Hum 40(4):825–838
3. Tomsovic K, Bakken DE, Venkatasubramanian V, Bose A (2005) Designing the next generation of real-time control, communication, and computations for large power systems. Proc IEEE 93(5):965–979
4. Liserre M, Sauter T, Hung J (2010) Future energy systems: Integrating renewable energy sources into the smart power grid through industrial electronics. IEEE Ind Electron Mag 4(1):18–37
5. Cao Y, Wu Q, Cheng S (1996) An improved Lyapunov function for power system stability analysis. Int J Control 65(5):791–802
6. Zhang Y, Bao Z, Cao Y, Li G, Chen G (2014) Long-term effect of different topology evolutions on blackouts in power grid. Int J Electr Power Energy Syst 62:718–726
7. Li Z, Shahidehpour M, Aminifar F, Alabdulwahab A, Al-Turki Y (2017) Networked microgrids for enhancing the power system resilience. Proc IEEE 105(7):1289–1310
8. Cyber-physical system. [Online]. Available: https://en.wikipedia.org/wiki/Cyber-physical_system
9. Cai Y, Huang T, Bompard E, Cao Y, Li Y (2017) Self-sustainable community of electricity prosumers in the emerging distribution system. IEEE Trans Smart Grid 8(5):2207–2216
10. Mets K, Verschueren T, Develder C, Vandoorn TL, Vandevelde L (2011) Integrated simulation of power and communication networks for smart grid applications. In: IEEE international workshop on computer aided modeling and design of communication links and networks

11. Yang CH, Zhabelova G, Yang CW, Vyatkin V (2013) Cosimulation environment for event-driven distributed controls of smart grid. IEEE Ind Inform 9(3):1423–1435
12. Georg H, Müller SC, Dorsch N, Rehtanz, C, Wietfeld C (2013) INSPIRE: integrated co-simulation of power and ICT systems for real-time evaluation. In: IEEE international conference on smart grid communications
13. Moradi-Pari E, Nasiriani N, Fallah YP, Famouri P (2014) Design, modeling, and simulation of on-demand communication mechanisms for cyber-physical energy systems. IEEE Ind Inform 10(4):2330–2339
14. Shi X, Xu Z, Li Y, Cao, Zhang C, Wen M, Liu F (2015) A hybrid simulation model for ICT-based wide-area damping control of power system. In: Proceedings of the 3rd international conference on industrial application engineering, pp 542–547
15. Nutaro J, Kuruganti PT, Miller L, Mullen S, Shankar M (2007) Integrated hybrid-simulation of electric power and communications systems. In: Proceedings of IEEE power engineering society general meeting, pp 1–8
16. Liberatore V, Al-Hammouri A (2011) Smart grid communication and cosimulation. In: Proceedings of IEEE energytech, pp 1–5
17. Lin H, Veda SS, Shukla SS, Mili L, Thorp J (2012) Geco: global event-driven co-simulation framework for interconnected power system and communication network. IEEE Trans Smart Grid 3(3):1444–1456
18. Godfrey T, Mullen S, Dugan RC, Rodine C, Griffith DW, Golmie N (2010) Modeling smart grid applications with co-simulation. In: Proceedings of IEEE 1st international conference on smart grid communications (SmartGridComm), pp 291–296
19. Levesque M, Xu D, Joos G, Maier M (2012) Communications and power distribution network co-simulation for multidisciplinary smart grid experimentations. In: Proceedings of the 45th annual simulation symposium. Society for Computer Simulation International, pp 1–7
20. Sun X, Chen Y, Liu J, Huang S (2014) A co-simulation platform for smart grid considering interaction between information and power systems. In: Proceedings of IEEE PES innovative smart grid technologies conference (ISGT), pp 1–6
21. Li W, Monti A (2010) Integrated simulation with VTB and OPNET for networked control and protection in power systems. In: Proceedings of the conference on grand challenges in modeling and simulation. Society for Modeling & Simulation International, pp 386–391
22. Li W, Monti A, Luo M, Dougal R (2011) VPNET: a co-simulation framework for analyzing communication channel effects on power systems. In: Proceedings of IEEE electric ship technologies symposium (ESTS). IEEE, pp 143–149
23. Li W, Luo M, Zhu L, Monti A, Ponci F (2013) A co-simulation method as an enabler for joint analysis and design of mas-based electrical power protection and communication. Simulation 89(7):790–809
24. Zhang Z, Huang X, Keune B, Cao Y, Li Y (2015) Modeling and simulation of data flow for vlan-based communication in substations. IEEE Syst J 99:1–12
25. Zhang Z, Huang X, He J, Yang Y, Cao Y (2013) Self-adaption packet-loss-based sampled value estimation algorithm and its error analysis. Autom Electr Power Syst 37(4):85–91 (in Chinese)
26. Yang Y, Huang X, Cao Y, Zhang Z, He J (2011) Security authentication for substation communication message and its real-time simulation. Autom Electr Power Syst 35(13):77–82 (in Chinese)
27. Hopkinson K, Wang X, Giovanini R, Thorp J, Birman K, Coury D (2006) Epochs: a platform for agent-based electric power and communication simulation built from commercial off-the-shelf components. IEEE Trans Power Syst 21(2):548–558
28. Zhu K, Chenine M, Lars Nordström (2011) ICT architecture impact on wide area monitoring and control systems' reliability. IEEE Trans Power Deliv 26(4):2801–2808
29. Li W, Li H, Monti A (2011) Using co-simulation method to analyze the communication delay impact in agent-based wide area power system stabilizing control. In: Proceedings of the grand challenges on modeling and simulation conference. Society for Modeling & Simulation International, pp 356–361

30. Hager U, Lehnhoff S, Rehtanz C, Wedde H (2009) Multi-agent system for coordinated control of facts devices. In: Proceedings of 15th international conference on intelligent system applications to power systems, pp 1–6
31. Terzija V, Valverde G, Cai D, Regulski P, Madani V, Fitch J (2010) Wide-area monitoring, protection, and control of future electric power networks. Proc IEEE 99(1):80–93
32. Hasan MS, Yu H, Carrington A, Yang T (2009) Co-simulation of wireless networked control systems over mobile ad hoc network using SIMULINK and OPNET. Communications IET 3(8):1297–1310
33. Rehtanz C (2003) Autonomous systems and intelligent agents in power system control and operation. Springer Science & Business Media. ISBN 3-540-40202-0
34. Li W, Ferdowsi M, Stevic M, Monti A, Ponci F (2014) Co-simulation for smart grid communications. IEEE Ind Inform 10(4):2374–2384
35. Panteli M, Kirschen D (2011) Assessing the effect of failures in the information and communication infrastructure on power system reliability. In: Proceedings of IEEE/PES power systems conference and exposition (PSCE), pp 1–7
36. Cai Y, Li Y, Cao Y, Li W, Zeng X (2017) Modeling and impact analysis of interdependent characteristics on cascading failures in smart grids. Int J Electr Power Energy Syst 89(Complete):106–114
37. Cai Y, Cao Y, Li Y, Huang T, Zhou B (2015) Cascading failure analysis considering interaction between power grids and communication networks. IEEE Trans Smart Grid 1–9
38. Junhua Z, Fushuan W, Yusheng X, Zhaoyang D (2011) Modeling analysis and control research framework of cyber physical power systems. Autom Electr Power Syst (in Chinese)
39. Schneider K, Liu C, Paul J (2006) Assessment of interactions between power and telecommunications infrastructures. IEEE Trans Power Syst 21(3):1123–1130
40. Xue B, Li N, Li S, Zhu Q (2010) Robust model predictive control for networked control systems with quantisation. Control Theor Appl IET 4(12):2896–2906
41. Martins Jota (2010) Design of networked control systems with explicit compensation for time-delay variations. IEEE Trans Syst Man Cybern C 40(3):308–318
42. Salo M, Tuusa H (2005) A new control system with a control delay compensation for a current-source active power filter. IEEE Trans Ind Electron 52(6):1616–1624
43. Nguyen N, Vanfretti L, Driesen J, Van Hertem D (2015) A quantitative method to determine ICT delay requirements for wide-area power system damping controllers. IEEE Trans Power Syst 30(4):2023–2030
44. Zhao J, Wen F, Xie Y, Li X, Dong Z (2010) Cyber physical power system: Architecture, implementation techniques and challenges. Autom Electr Power Syst 34(16):1–6 (in Chinese)
45. Wan J, Yan H, Li D, Zhou K, Zeng L (2013) Cyber-physical systems for optimal energy management scheme of autonomous electric vehicle. Comput J 56(8):947–956
46. Banerjee A, Venkatasubramanian K, Mukherjee T, Gupta S (2012) Ensuring safety, security, and sustainability of mission-critical cyber–physical systems. Proc IEEE 100(1):283–299
47. Tan Y, Li Y, Cao Y, Shahidehpour M (2017) Cyber-attack on overloading multiple lines: a bilevel mixed-integer linear programming model. IEEE Trans Smart Grid (99):1–1
48. Creery A, Byres E (2005) Industrial cybersecurity for power system and SCADA networks. In: Petroleum and chemical industry conference. IEEE
49. Ten C, Liu C, Govindarasu M (2007) Vulnerability assessment of cybersecurity for SCADA systems using attack trees. In: Power engineering society general meeting. IEEE
50. Liu X, Shahidehpour M, Li Z, Liu X, Cao Y, Li Z (2017) Power system risk assessment in cyber attacks considering the role of protection systems. IEEE Trans Smart Grid 8(2):572–580

Chapter 2
Modeling and Analysis Techniques of Interdependent Network

Smart grid is a revolutionary artificial system with many advanced technologies for monitoring and controlling, e.g., advanced modern sensor and measurement technology, communication and information technology [1]. For the highly safe and economic operation and control of the smart grid, the comprehensive information system containing numerous measurements, monitoring, and management information is preferred. However, smart grid may suffer threats from communication systems, e.g., hackers and viruses. Some communication functions might be infected and disabled in some extreme cases, such as the transformer exploding due to an intentional attack, which are harmful to the implementation of control for power grid even causing the cascading failure [2]. The interdependent relationship between the power grid and the information network further increases the complexity of the safety assessment in power systems [3]. Therefore, the interacting mechanism including the model of interdependent networks should be investigated to analyze and solve these problems, thus to avoid the occurrence of cascading failure.

2.1 Overview of Cascading Failure in Interdependent Network

Several large blackouts have been occurred in a few countries due to the cascading failures triggered by various possible threats, for example, the Italy blackout in 2003; the China Hainan blackout in 2005; China South Power Grid blackout in 2008; the Brazil blackout in 2009; and the India blackouts in 2012.

Several complex network theories are considered to analyze the mechanism of cascading failure, which can be summarized as follows:

(1) the frequency of large blackouts is governed by a power law which makes the risk of large blackouts. Many works have found that the probability distribution of the scale of blackout somehow changes from the exponential tail form to the power low form due to the increasing load [4], unbalance power flow [5], etc.;

Y. Cao et al., *Cyber-Physical Energy and Power Systems*,
https://doi.org/10.1007/978-981-15-0062-6_2

(2) overall load or stress relative to operating limits is a key factor affecting the risk of cascading failure [6].

Cascading failures due to the interactions have been observed several times during the recent decades [7, 8]. In 2003, some communication nodes lost functions owning to some power stations' shutdown. Because of the lack of information, more power stations went into blackout which finally led to a catastrophe in Italy [9]. A similar accident within Northeast American in 2003 was essentially caused by a software bug.

Many approaches are employed to analyze the interactions among different networks. In order to describe the dynamic behaviors of cascading failures propagating among two different networks, various models, such as agent, statistic or graph theory-based models, are studied. During the beginning period of the theoretical study of interaction, it was concluded that a scale-free network was highly robust to random attacks, and two coupled scale-free networks were sensitive to random attacks [9]. Currently, several research studies have presented different analytic descriptions of interdependencies. Reference [8] has built an interactive model to raise the importance of the interdependency between power grids and the SCADA from a security perspective. Improved topological parameters are proposed to identify the vulnerability of power systems [10, 11]. References [12, 13] have proposed mathematical models based on the Petri net to explain the interaction that would finally cause a catastrophe. From the view of cyber security, some research studies have focused on the impacts of incomplete information and operating margins [13–15]. Dynamic simulations for cyber-physical systems have been proposed to match the performance of power grids under different cyber-attacks [16, 17].

However, the interaction represents a bidirectional relationship between power systems and the dispatching data network in several countries especially in China. Few researches have begun to address this issue although focuses are only on the structural and transmission characteristics of dispatching data networks [18, 19]. Therefore, a modeling approach should be considered to investigate the interactions between the power system and the dispatching data network.

2.2 Modeling for Interdependent Network

In terms of the blackout occurred in Italy, a general model was proposed to describe the cascading failure propagating between the two interacting networks [7]. The idea to model interdependent networks is two networks, A and B, with the same number of nodes N, the functioning of node A_i ($i = 1, 2, \ldots, N$), in network A, depends on the ability of node B_i in network B, to supply a critical resource, which means if node A_i stops functioning, node B_i stops functioning, and vice versa.

Here, we take a 6 nodes system as an example, system A and system B with the same number nodes, which defines a one-to-one correspondence between nodes of network A and B. A fault was intrigued by attacking any node within network A, and the stages of cascading failure are given as followed.

(1) Attacking the node within network A randomly, and the node could not work normally;
(2) *Stage* 1: The attacked node and the lines linked by the node are firstly removed. Accordingly, due to the correspondence between the two networks, the nodes and the lines within the network B are also removed. Then the network A is divided into three "a_1 clusters," i.e., a_{11}, a_{12}, a_{13}, a_{11}, a_{12} are islands;
(3) *Stage* 2: Due to a_{11}, a_{12} are islands, the nodes could not work normally. We define the b_2-sets as the sets of B-nodes that are connected to a_1-clusters by $A \leftrightarrow B$ links, so the interdependent nodes in the network B will break down too, i.e., the loss of b_{21}, b_{22}. Then all the b_2-links connecting different b_1 sets must be removed. Then, the network B was divided into four b_2-clusters, i.e., b_{21}, b_{22}, b_{23} and b_{24}, b_{21}, b_{22} are islands;
(4) *Stage* 3: In the network A, the islands a_{31}, $a_{32,}$ and a_{33} are the nodes directly connected to islands b_{21} and b_{22} in the network B. According to the interdependency between the two systems, network A was divided into four a_3-clusters which establish a one-to-one correspondence between a_3- and b_2-clusters. The system tends to be stable, and the cascading failure will be terminated (Fig. 2.1).

From the above description, the communication nodes in the information networks are interdependent on physical nodes in the power network. Any node's failure in any network will cause the corresponding node fails in another network; however, this idea is not always suitable for the real power system: Generally, the failure of communication nodes will not completely affect the normal operation of the corresponding power nodes. At the same time, the energy flow in power system and information network needs to satisfy their respective characteristics. The differences

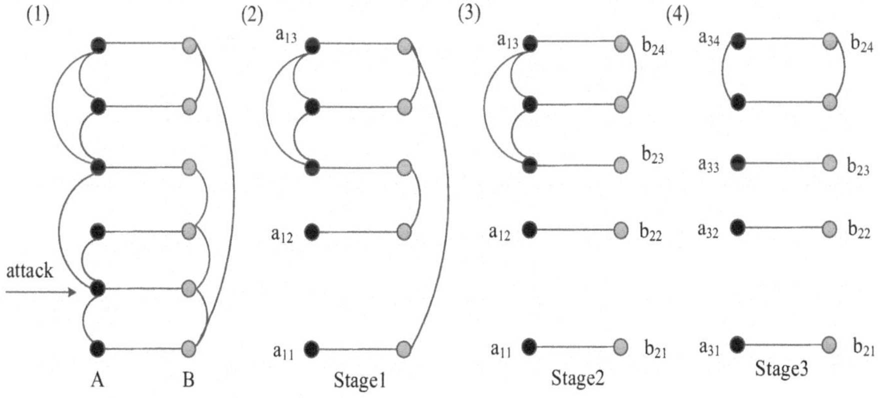

Fig. 2.1 General interaction model of coupling networks

are, for example, the node injection current in the power system should obey Kirchhoff's law, and the data packet transmission in information network is carried out according to certain routing strategies, etc.

Power systems are characterized by several physical and operational constraints. The system operators ensure system stability and safety by those constraints through control and monitoring systems (SCADA, RTU, etc.). A huge number of interactive devices (generators, loads, and new electrical devices) located at various spatial locations with different functionalities cross all voltage levels. So, the interaction between power grid and information grid becomes more closely, and the self-recognition of grids is highly depended on massive amount of accurate information from the dispatching data network. Similarly, the control and monitoring systems are also complex, and their behaviors are closely coupled. To acquire the overall information of power grids, power grids require integrated, efficient, and reliable information and communication systems (ICTs) [4].

A stronger dispatching data network can enhance the capability of operators to monitor and control the entire system. However, in some cases, cyber-attacks on switch nodes can cause information loss or affect protection appliances. In these cases, the operators sometimes cannot make correct decisions or any decisions due to lack of information or specific indices to access overall performances.

The interdependent systems can be decomposed into two general layers, the physical layer and cyber layer, as shown in Fig. 2.2. The physical layer includes electrical devices for generation, transmission, distribution, and consumption. The cyber layer is represented by the dispatching data network, which is used to gather, transfer,

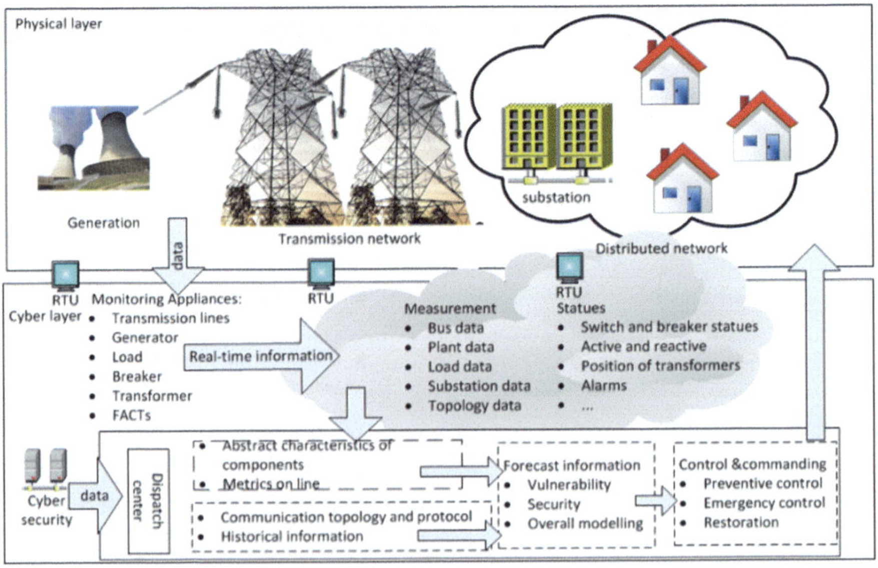

Fig. 2.2 Framework for power system operation and control

and process information with the consideration of the structural and transmission characteristics [18, 19]. Between these two layers, numerous data of all devices are sent to the dispatch center via the dispatching data network. The dispatch center thus can assess the system states. Additionally, the dispatch center can send back control commands to power grids via the dispatching data network.

The dispatching data network works as the fundamental infrastructure to support the operation of power systems, especially when failures occur in power grids. However, because of the interdependency, failures in power grids might bring failures to dispatching data networks. Under such an emergency, first, the abnormal message packets should be sent to the dispatch center quickly. Second, the dispatch center will provide effective control, which is an essential step in preventing cascading failures.

2.3 Model for Communication Network

2.3.1 Complex Network Background

The network can be represented by a graph composed of *vertices* (buses in power grids, nodes in the dispatching data network) and *edges* (branches in power grids, links in the dispatching data network). The edges can be described as *weighted/unweighted* and *directed/undirected.*

In an undirected graph with n vertices, the degree of vertex i is the number of edges connected to the vertex, as in (2.1), where if vertices i and j are connected, $a_{ij} = 1$; else, $a_{ij} = 0$.

$$K_i = \sum_{j}^{n} a_{ij} \tag{2.1}$$

The *characteristic path length* of the entire network is defined by (2.2), where d_{ij} is the *shortest path* connecting vertices i and j; the *length* of a path between vertices i and j is the number of edges in the path.

$$L = \frac{1}{n(n-1)} \sum_{i \neq j} d_{ij} \tag{2.2}$$

The betweenness is an indicator of a node's centrality in a network. However, a rather high complexity is imposed on the betweenness computation, because of most of real-world large networks displaying a hierarchical community structure. We proposed a new hierarchical decomposition approach to speed up the betweenness computation for complex networks [20]; but for convenience of analysis and calculation, here we use the basic definition to calculate the betweenness.

The betweenness is the number of shortest paths which through a vertex or edge. The betweenness of node v can be formatted as in (2.3). In (2.3), $\sigma_{ij(\mathrm{v})}$ is the number of the shortest path between nodes i and j through node v, and σ_{ij} represents the total shortest path between nodes i and j. A higher value of betweenness of a node implies the shortest path through it; thus, a critical node of the network can be identified by ranking betweenness [10, 21].

$$B(v) = \sum_{i} \sum_{j} \frac{\sigma_{ij}(\mathrm{v})}{\sigma_{ij}} \tag{2.3}$$

In the context of complex networks, there are three types of network, i.e., random network [22], small-world network [23], and scale-free network [24]. The random network (also known as ER network) is based on random graph theory, which is used to describe the communication network and biological network. The clustering coefficient and average shortest distance are small. Comparatively, the small-world network is originated from random networks, which has high clustering coefficient and short average shortest distance. In contrast, the power law distribution is the typical characteristic of a scale-free network (also known as BA network). The BA network is robust to the random attack but sensitive to intentional attacks to hob nodes.

When considering the structural security of a network, an approach is employed to analyze the consequences after removing a group of vertices or edges and to find out what would happen in cascading failures after load redistribution. The damage caused by cascading failures is quantified as the relative size of the largest connected component $G_{\max} = N'/N$. N and N' are the number of vertices or edges in the largest connected component before and after cascading failures.

2.3.2 *Topological Models of Communication Network*

Minor failures, such as incorrect actions of protections, data lost or delayed in the dispatching data network, may cause a chain of components tripping, which is usually accompanied by the phenomenon of voltage and frequency collapse [14, 25]. In fact, the interaction of different systems also extends to the area of faults: a small disturbance not only spreads within the network but also causes unpredictable influences on other systems [26]. In this section, we present a model to describe the interaction between power grids and dispatching data network.

According to the complex networks theory mentioned in Sect. 2.3.1, both the power system and the dispatching data network can be modeled as $G = (V, L)$, where V is a set of vertices, L is a set of edges. Generally, there are two types of the dispatching data network in China: double-star network and mesh network [18, 19]. Figures 2.3 and 2.4 show these two types for the IEEE 39-bus system and China's Guangdong 500 kV system.

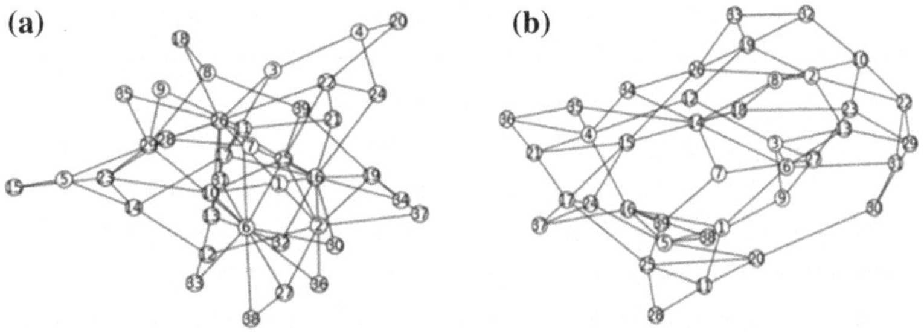

Fig. 2.3 IEEE 39-bus system dispatching data network. **a** Double-star structure; **b** mesh structure

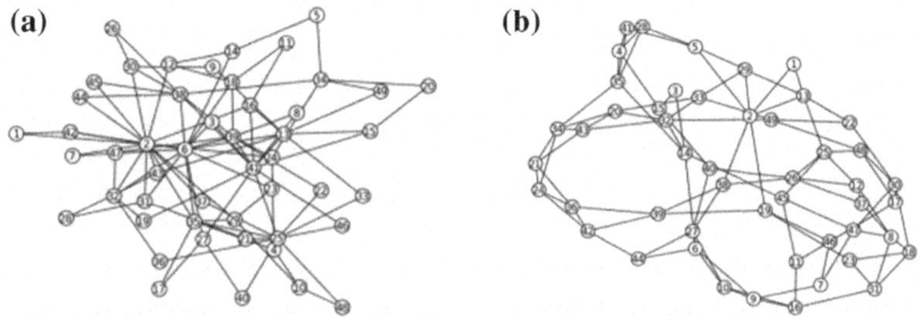

Fig. 2.4 China's Guangdong 500 kV system power dispatching data network. **a** Double-star structure; **b** mesh structure

Figures 2.3a and 2.4a are the double-star dispatching data networks which are also the scale-free networks. The main characteristic of the scale-free network is that there are few hub nodes with more neighbors. The dispatch center is represented as one of the hub nodes. Compared with the double-star network, mesh dispatching data networks are the small-world networks as shown in Figs. 2.3b and 2.4b, and distributions of degree and betweenness are much more balanced.

Table 2.1 shows structural characteristics of these two networks, and Table 2.2 shows the descending orders of degree and betweenness of these two networks. N and M are the numbers of vertices and edges. K and $\bar{k}$ are the degree of a vertex

Table 2.1 Structural characteristics of two dispatching data networks

	N	M	$\bar{k}$	C	C_{random}	L	L_{random}
(a) IEEE 39-bus system dispatching data network							
Double-star	39	74	3.794	0.143	0.104	2.433	2.639
Mesh	39	78	4.0	0.208	0.061	2.892	2.678
(b) China's Guangdong 500 kV system dispatching data network							
Double-star	49	94	3.836	0.169	0.053	2.582	2.834
Mesh	49	98	4.0	0.163	0.082	3.107	2.891

Table 2.2 Descending orders of degree and betweenness

Double-star		Mesh		Double-star		Mesh	
Node	K	Node	K	Node	B	Node	B
(a) IEEE 39-bus system dispatching data network							
16	11	16	6	16	162	15	87
26	10	14	6	26	154	14	84
6	10	4	6	6	126	6	80
(b) China's Guangdong 500 kV system dispatching data network							
2	20	2	7	2	488	6	133
6	15	27	6	6	315	2	109
41	9	14	6	25	141	27	98

and the average degree of the network, respectively. C and C_{random} are clustering coefficients, where C_{random} is calculated for random networks with the same size. L and L_{random} are the average shortest length, where L_{random} is calculated for random networks with the same size.

As shown in Table 2.1, for the mesh network, $L > L_{\text{random}}$ and $C > C_{\text{random}}$ which are the main characteristics of small-world networks. The dense relation (L) is helpful to improve the efficiency of transmission [18]. Because of the smaller L, double-star networks have higher transmission efficiency than mesh ones; C describes the aggregate level of nodes. The higher C is because more communication nodes are clustered around hub nodes. In double-star networks, the hub node makes the network vulnerable to intentional attacks [24].

Moreover, from Table 2.2, compared with $\bar{k}$, the degrees of hub nodes in the double-star network are much larger than the degrees of others, which is a key feature of scale-free networks. The degree distribution of the mesh structure is much more balanced. The betweenness reflects the amount of transferred data in an indirect way. The betweenness of hub nodes in the double-star network is also higher than that of other nodes; however, it exhibits more similarities to all nodes in the mesh network. The main reason is that, in the mesh structure, there are more transmission links among nodes.

Reference [27] notes that interdependencies between two networks can be defined as inter-similarities from topological views. As critical switch nodes have more links to other nodes, the degree is a suitable index to identify critical nodes in the dispatching data network. For power grids, both degree and betweenness are used to identify critical nodes briefly. Because double-star networks show scale-free characteristics, the one-to-one interdependency of nodes between the dispatching data network and the power grid is "degree to degree." For example, the node with the largest degree in the dispatching data network is coupled with the node with the largest degree in the power grid. Because mesh networks show small-world features, the one-to-one interdependency is "degree to betweenness" as in (2.4), which is similar to [27]. That is, the vertex with the largest degree in the dispatching data network is coupled with the node with the largest betweenness in the power grid.

$$r = \sum_{BK} BK(e_{BK} - P_B P_K) \tag{2.4}$$

where r is the correlation of a pair of depending nodes in the two networks; B and K are the betweenness and degree; e_{BK} is the joint probability of a coupling link connected to the node with the betweenness B and the degree K; P_B and P_K are the betweenness distribution and degree distribution, respectively.

2.3.3 Information Network Routing Strategy

All operating parameters associated with transmission lines are sent to the dispatch center step-by-step via the dispatching data network. We always assume that the capacity of communication node is enough to handle all information. We do not consider the method of gathering and dealing with the information in the dispatching data network.

The communication node in dispatching data networks is coupled with the physical bus in power grids. Thus, for each transmission line, there are two related communication nodes. In the normal state, we assume that only the communication node coupled with the high voltage side of the transmission line is used to exchange the information.

At every step, a message packet can be received and sent out only once by each communication node. The rules for data exchange are as follows: At the beginning, the source node produces message packets. At the next time step, if the destination node is in its neighbor sets, the message packets are sent to the destination node directly, and the exchange ends; otherwise, the message packets will be transferred to one of the neighbor nodes based on the chosen probability P, as in (2.10) [28]. Finally, the chosen neighbor node becomes the new source node. The new source node will repeat the behaviors at the next time step until the message packets are sent to the destination node. Sending out the message packets from the source nodes follows the rule of first-in-first-out, and it avoids the message packets to be transferred in the same edge.

The power information network adopts Open Shortest Path First (OSPF) routing protocol, which is mainly based on the "shortest path transmission" routing strategy. When the shortest path contains some hub nodes with larger degrees, the queueing time of packets at the nodes may extend, due to the larger load of the hub nodes. Therefore, when selecting the routing strategy, the packet queue length of each node should be taken into consideration.

For a source node i and its neighbor sets $L_i, j \in L_i$, the chosen probability P_j is defined as in (2.5), where d_j is the length of the shortest path between node j and the destination node; c_j is the number of message packets in the queues of node j; and h_d and β are constants.

$$\begin{cases} H_j = h_d d_j + (1 - h_d)c_j \\ P_j = \frac{e^{-\beta H_j}}{\sum\limits_{m \in L_i} e^{-\beta H_m}} \end{cases} \quad (2.5)$$

2.4 Analysis of Blackout Caused by Interdependent Network

2.4.1 *Cascading Failure Analysis Based on Interdependent Network*

As we know the interaction between power grid and the dispatching data network, i.e., the interdependent network, will exert a tremendous impact on proper operation in the dispatching center. We establish a more accurate interdependent network model, with several faults circumstances taken into consideration.

Scenario 1: Potential faults occur in generator or relay devices caused by various grid component, and the state information of transmission lines is transmitted to the control center through communication node in high voltage side.

Scenario 2: Malfunction of transformer branch in high voltage terminal and generator bus of power network occur, and the corresponding communication nodes fail. When other branches of the power grid fail, the corresponding communication nodes could not work normally with probability P_1.

Scenario 3: The communication node is hub node where the fault occurred, all the nodes in the neighborhood fail, while the communication node is not hub node, the nodes in the neighborhood area fail in probability P_2, P_2 is a constant.

Scenario 4: When a line ij is overloaded at time t, we calculate the time t_{ij} of inverse-time overcurrent protection as in (2.7) [29], where $K = 7$, $\alpha = 0.3$, I_{ij} and $I_{\text{set}ij}$ are the current and setting current, and $t = t + 1$; if $t_{ij} < t$, the protection acts; thus, the overloaded line is tripped. If complete control is finished before the overloaded line is tripped, we define it as an effective control; otherwise, the preventive control or corrective control commands are not received, and thus the cascading failures occur.

$$t_{ij} = \frac{K}{\left|L_{ij}/I_{\text{set}ij}\right|^{a} - 1} \quad (2.6)$$

A simple case is used to explain the interactions, as shown in Fig. 2.5. The power grid A and the dispatching data network B have the same number of nodes. Each node in A_i depends on B_i, as described in Section IV-A. If a transmission line A_{ij} is overloaded or out of service, the abnormal message packets of A_{ij} are produced by B_i. The hidden failures in both A and B are considered as follows: the outage of A_{ij}

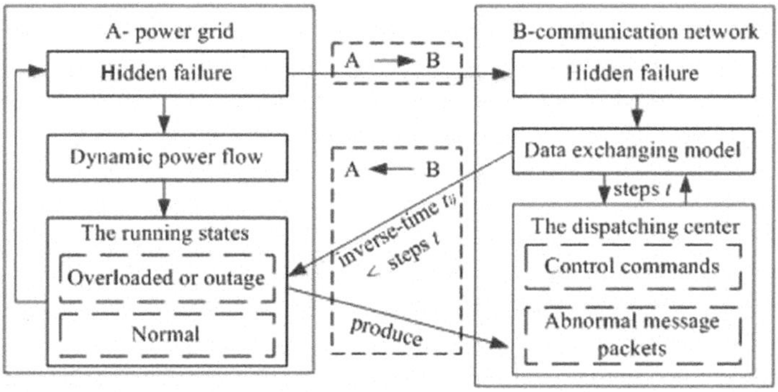

Fig. 2.5 Formal model of interaction

may cause the hidden failures of its neighbors with the low probability P_1; if B_i is out of service, its neighbors have risks of being out of service with P_1.

The influence of A → B: The tripped transmission lines A_{ij} are removed from network A. If A_i is out of service, B_i has a risk of working abnormally [10]. If A_i is the transformer's high voltage side or generator bus, B_i is exposed to the incorrect state. Otherwise, B_i will be tripped with a low probability of P_2.

The influence of B → A: The structure of B determines the steps needed for complete control. The abnormal message packets from B_i are sent to the dispatch center step-by-step. The dispatch center responds to the failures if it has received the abnormal message packets. The control commands are also sent back step-by-step. If $t_{ij} < t$, the overloaded line is tripped.

2.4.2 *Dynamic Power Flow in Power System*

Taking the dependencies of load on voltage, frequency, and generator regulation characteristics into consideration [30], we use dynamic power flow calculation to redistribute power flow during disturbances.

(1) Load characteristic: According to the voltage and frequency of bus i, the load of bus i is presented as in (2.7) [31], where P_{di} and Q_{di} are the active and reactive load, respectively. P_{d0i} and Q_{d0i} are the active and reactive load at the rated voltage and frequency, respectively. Δf is the variation of the frequency. V_i is the voltage magnitude. And k_i, a_i, b_i, c_i, a_i', b_i', c_i' are constants, where $a_i + b_i + c_i = 1$ and $a_i' + b_i' + c_i' = 1$.

$$
\begin{aligned}
P_{di} &= P_{d0i}(1 + k_i \Delta f)(a_i + b_i V_i + c_i V_i^2) \\
Q_{di} &= Q_{d0i}(1 + k_i' \Delta f)(a_i' + b_i' V_i + c_i' V_i^2)
\end{aligned}
\tag{2.7}
$$

(2) Generator governor characteristics: Generation can be adjusted autonomously when some disturbances exist in power systems, as in (2.8), where in bus i, P_{gik} is the generation of generator K, $P_{\mathrm{set}ik}$ is the rated active power of generator K, and K_{gik} is the unit power regulation of generator K.

$$\begin{aligned} P_{gi} &= \sum_{k=1}^{N_{gi}} P_{gik} = \sum_{k=1}^{N_{gi}} (P_{\mathrm{set}ik} - K_{gik}\Delta f) \\ P_{\min ik} &\le P_{gi} \le P_{\max ik} \end{aligned} \tag{2.8}$$

If the load changes by ΔP_{D0}, when only considering the load frequency characteristics, the change of load is ΔP_D, and the change of generation is ΔP_G. The variation of generation and load can be expressed as in (2.9)

$$\begin{aligned} \Delta P_{D0} + \Delta P_D &= \Delta P_{D0} + k_i \Delta f = \Delta P_G \\ &= -K_g \Delta f \to \Delta P_{D0} = -(K_g + k_i)\Delta f \end{aligned} \tag{2.9}$$

(3) Power flow equations: The active and reactive power load flow equations can be written as in (2.10) and (2.11). The unbalanced active power of each PQ bus or PV bus is as in (2.10); the unbalanced reactive power of each PQ bus is as in (2.11)

$$\begin{aligned} \Delta P_i &= P_i - (P_{gi} - P_{di}) \\ &= V_i \sum_{J=1}^{n} V_j (G_{ij}\cos\delta_{ij} + B_{ij}\sin\delta_{ij}) - \sum_{k=1}^{N_{gi}} (P_{\mathrm{set}ik} - K_{gik}\Delta f) \\ &\quad + P_{d0i}(1 + k_i\Delta f)(a_i + b_i V_i + c_i V_i^2) \\ &= \sum_{j=1}^{n} |V_i||V_j||Y_{ij}|\cos(\theta_{ij} + \delta_j - \delta_i) - \sum_{k=1}^{N_{gi}} (P_{\mathrm{set}ik} - K_{gik}\Delta f) \\ &\quad + P_{d0i}(1 + k_i\Delta f)(a_i + b_i V_i + c_i V_i^2) \end{aligned} \tag{2.10}$$

$$\begin{aligned} \Delta Q_i &= Q_i - (Q_{gi} - Q_{di}) \\ &= V_i \sum_{J=1}^{n} V_j (G_{ij}\sin\delta_{ij} - B_{ij}\cos\delta_{ij}) \\ &\quad + Q_{d0i}(1 + k_i'\Delta f)(a_i' + b_i' V_i + c_i' V_i^2) \\ &= \sum_{j=1}^{n} |V_i||V_j||Y_{ij}|\sin(\theta_{ij} + \delta_j - \delta_i) \end{aligned}$$

$$+ Q_{d0i}(1 + k_i' \Delta f)(a_i' + b_i' V_i + c_i' V_i^2) \quad (2.11)$$

(4) The dynamic power flows are calculated when some components lose functionality during the cascading process. The simulation model can be expressed as follows:

Step 1: initialization: get information on the load, generators, and all parameters of buses and branches; compute the power flows by the Newton model;
Step 2: attacking: randomly remove a branch in the power grids; then, modify the nodal admittance matrix; compute the power flows by the Newton model;
Step 3: load redistribution: if any branch is overloaded, proceed to the next step; otherwise, end the simulation;
Step 4: frequency stability: calculate ΔP and Δf according to (2.9). If Δf is within limits, compute Δf and proceed to the next step; otherwise, end the simulation and send warnings of instability;
Step 5: beginning of the iteration: specify voltage magnitudes and phase angles;
Step 6: solve equations: calculate the active and reactive unbalanced ΔP_i and ΔQ_i according to (2.10) and (2.11);
Step 7: convergence: if the deviations of both the active and reactive power satisfy the convergence criteria, proceed to the next step; otherwise, go back to step 6;
Step 8: report: get the power flows;
Step 9: cascading process: if any branches are overloaded, remove them, modify the nodal admittance matrix, compute the power flows by the Newton model, and then go to step 4; otherwise, end the simulation.

2.4.3 Cascading Failure Simulation

An N-node small-world network is built to represent the mesh dispatching data network. The first step is to build an N-node regular graph where each node is connected to $k = 4$ nearest neighbors. Then, for each edge, rewire it with probability $p = 0.3$ [23]. To represent the double-star dispatching data network, we develop an N-node scale-free network. Firstly, the original graph has $m_0 = 2$ nodes, and then $m = 2$ new nodes are added at each step until the size of the graph reaches N nodes. Each new node is preferentially attached by an edge to the existing node with higher degree [24].

According to [27] and (2.4), Table 2.3 shows the values of two types of inter-similarities, "degree to degree" for power grids coupled with double-star dispatching data networks and "betweenness to degree" for those coupled with mesh dispatching data networks. More inter-similarities indicate higher robustness against random failures [27].

Table 2.3 Inter-similarities of power grids and dispatching data networks

Power grids	Double-star	Mesh
IEEE 39-bus system	0.1798	0.1135
Guangdong 500 kV system	0.1333	0.04164

Thus, considering the interdependency, the approach of the cascading process simulation is summarized as follows:

Step 1: initialization of power systems and dispatching data networks: give $t = 0$, in (2.5), $h_d = 1$, $\beta = 10$;
Step 2: triggering: remove a transmission line in the power grid; produce outage message packets from the coupled communication node;
Step 3: hidden failures: search failure lines in the power grid and failure communication nodes due to hiding failures; produce the outage message packets from the coupled communication node if failures exist in power grids;
Step 4: data exchange: $t = t + 1$, all of the message packets are delivered via the data exchange model in Sect. 2.3.3;
Step 5: identify status: calculate the power flow; calculate the inverse-time t_{ij}, of overloaded lines; end the simulation if all transmission lines are running under permission;
Step 6: interaction: trip the overloaded lines if $t_{ij} < t$; produce the outage message packets from the coupled communication node if failures exist in power grids;
Step 7: redoing hidden failures: if some transmission lines are tripped, repeat step 3 and then proceed to the next step;
Step 8: locating the message packets: produce some automatic optimal control commands if the dispatch center has received message packets; if not, return to step 4;
Step 9: locating the control commands: return to step 4 if the commands have not arrived; otherwise, end the simulation.

2.5 Case Studies

The simulation model is built by means of Python 2.6.5. According to the cascading simulation mentioned in section IV-E, we calculate the size of the largest connected components and the load shedding of power grids in each simulation. We consider hidden failures in both power grids and dispatching data networks and then update each topological structure. The dynamic power flow calculation is used for load redistribution. If the dispatch center receives abnormal message packets, we consider two strategies to resist cascading failures: the reactance x_{ij} adjustment of overloaded lines by FACTS, as in (2.12) [32], and LP re-dispatch to shed some loads.

$$x_{ij} = x_{ij0} * L_{ij}/S_{ij}, S_{ij} < L_{ij} < S_{\max ij} \tag{2.12}$$

where L_{ij} is the power flow of overloaded line ij; S_{ij} and $S_{\max,ij}$ are the capacity of the overloaded line ij and its thermal limit.

Figure 2.6 reports the probability of load shedding under a random attack in the IEEE 39-bus system and China's Guangdong 500 kV power system. Each run simulates 100 cascades. The results show the characteristics of the power tails.

The probability of blackout is lower when the power grid is coupled with a double-star structure network. Thus, the resistance to random attacks when using the double-star structure is higher than when using the mesh one. The probability of blackout is lower when the power grid is coupled with a double-star structure network. Thus, the resistance to random attacks when using the double-star structure is higher than when using the mesh one. Thus, the resistance to random attacks when using the double-star structure is higher than when using the mesh one. From the topological view, because the power grid is a small-world network, the propagation of failures is

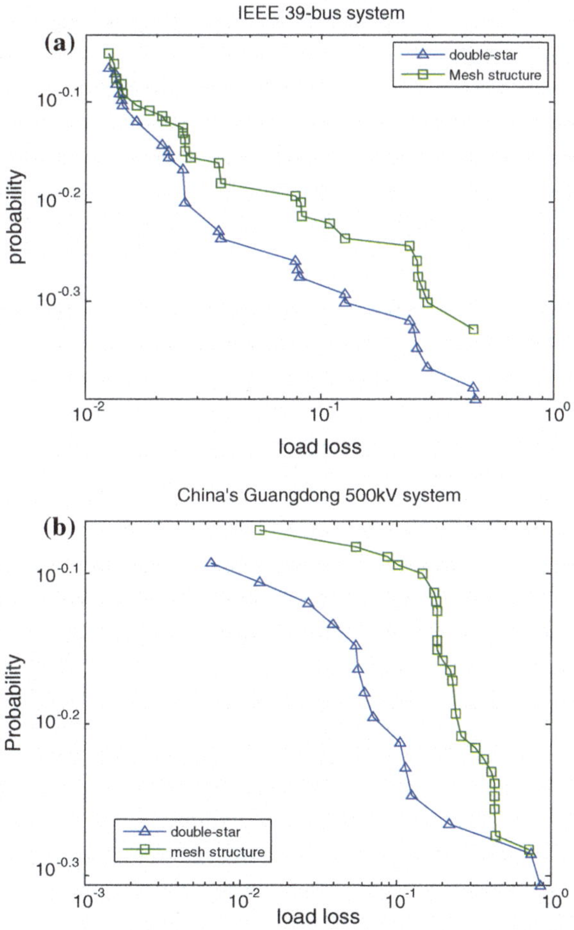

Fig. 2.6 Log–log plots of blackout distributions. **a** IEEE 39-bus system; **b** China's Guangdong 500 kV system

easier and faster under random attacks, which correspondingly affects the efficiency of dispatching data networks. The mesh dispatching data network is also a small-world network which is sensitive to random failures; thus, the interdependencies between power systems and the mesh dispatching data network extend the area of cascading failures. However, the double-star network is a scale-free one, which is robust to random failures, and thus using the double-star structure dispatching data network for the power system is better in case of random attacks.

We show the result of intentional attacks on power grids in Fig. 2.6. The transmission lines of higher betweenness are much more important because their power flows are higher (more transmission corridors pass through them). The abnormal message packets of the higher betweenness transmission lines should be prior to transmission to the dispatch center. Information of higher betweenness lines is sent by hub nodes with the interdependency of "betweenness-degree" (the power grid coupled with the mesh dispatching data network). Thus, when the line of higher betweenness is removed, the abnormal information can be sent to the dispatching center quicker so that the consequence is less serious than when the interdependencies are "degree to degree." Moreover, coupling with mesh networks, its largest connection is always above 60%, which is higher than that of a double-star network because the double-star network is the scale-free network and is fragile to intentional attacks.

Furthermore, index 21 in Fig. 2.7a represents the line (19, 16) of higher betweenness (betweenness = 140, ranking 8th in the descending orders of edges' betweenness). The outage of (19, 16) leads to isolating buses 19, 20, 33, and 34. Both buses 33 and 34 are generator buses, which have the lowest degrees and betweenness. The rankings of the degrees of buses 19 and 20 are 9th and 24th, respectively, and the rankings of their betweenness are 9th and 23rd, respectively. Although the rankings of both degree and betweenness in power grids are nearly the same, the degrees of the coupled communication nodes 19, 20, 33, and 34 in the double-star network are respectively smaller than their degrees in the mesh network because the double-star network exhibits the scale-free features. Specifically, in the double-star network, the degrees of communication nodes 19, 20, 33, and 34 are 5, 2, 2, and 2, respectively, whereas they are 5, 4, 3, and 3 in the mesh network, respectively. The higher degree is helpful for sending abnormal message packets to the dispatch center, as well as for implementing the control. Thus, the load shedding is lower when power grids are coupled with mesh networks. Observing the cascading process triggered by removing the medium-betweenness line, such as (26, 25) indexed 29 in Fig. 2.7a, (betweenness = 100, ranking 14th), but node 26 has the highest degree in power grids, which means that node 26 is the hub node in the double-star dispatching data network. Under "degree to degree," the outage messages can be sent to the dispatch center quickly, as well as for implementing the control. Thus, for the interdependent networks, new indices are necessary to access the critical lines in power grids to improve the robustness against failures.

In practice, power systems are faced with high probabilities of random attacks, such as natural disasters and hidden failures; thus, the double-star dispatching data network is much more suitable for power systems when considering interdependency.

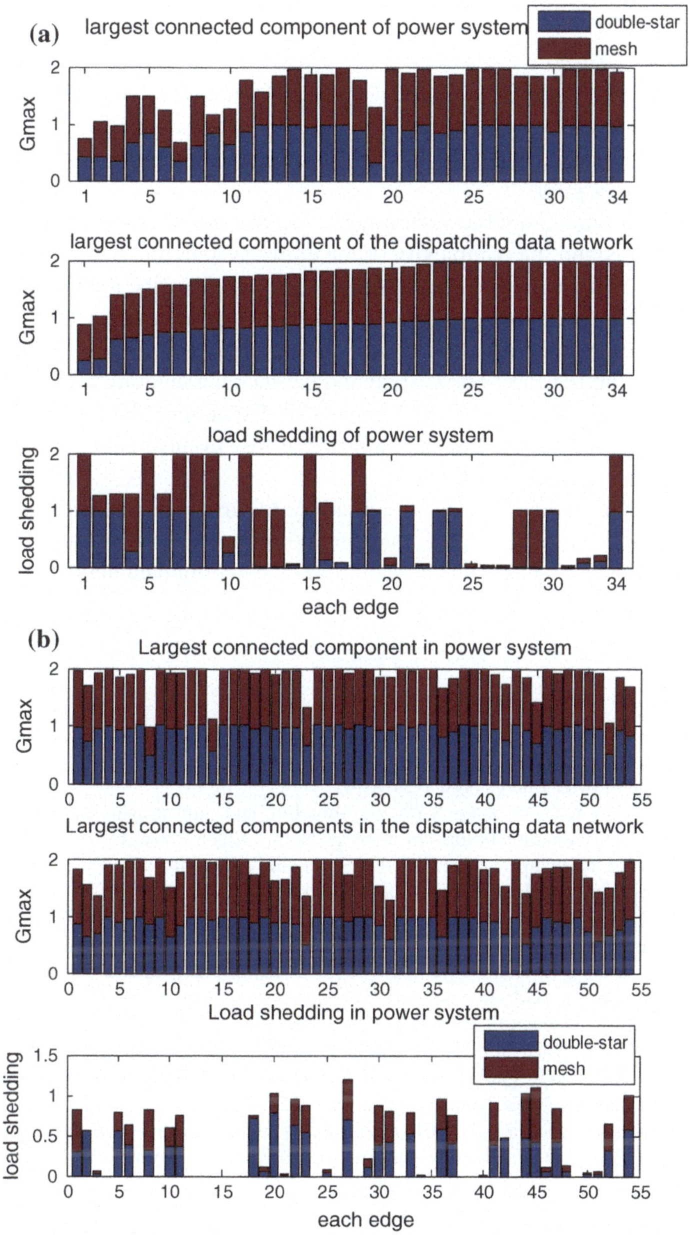

Fig. 2.7 Intentional attack on each edge of the power system. **a** IEEE 39-bus system; **b** China's Guangdong 500 kV system

2.6 Summary

The increasing interdependence between power systems and dispatching data networks makes the security problem more complex. Interdependency can accelerate the cascading process; thus, capturing the interaction unsuccessfully may lead to blackouts of power grids.

Modeling the interaction between power systems and dispatching data networks is a way to understand the complexity of the entire system and improve the security and reliability. This chapter gives an approximation to describe the interdependence based on the dynamic power flow model. The topological and partial transmission characteristics of dispatching data networks are considered. By using the proposed model, we take the IEEE 39-bus system and China's Guangdong 500 kV system as examples to investigate the structural imparts of dispatching data networks on load shedding in case of different attacks on power grids. In most cases, the double-star structure is better than the mesh one.

To improve the ability to prevent cascading failures, the development of smart grids should consider the structural characteristics of dispatching data networks and the interdependence between power grids and dispatching data networks.

References

1. Shi X, Li Y, Cao Y, Tan Y (2015) Cyber-physical electrical energy systems: challenges and issues. CSEE J Power Energy Syst 1(2):36–42
2. Cai Y, Li Y, Cao Y, Li W, Zeng X (2017) Modeling and impact analysis of interdependent characteristics on cascading failures in smart grids. Int J Electr Power Energy Syst 89:106–114
3. Cai Y, Cao Y, Li Y, Huang T, Zhou B (2015) Cascading failure analysis considering interaction between power grids and communication networks. IEEE Trans Smart Grid 7(1):530–538
4. Yan J, Zhu Y, He HB, Sun Y (2013) Multi-contingency cascading analysis of smart grid based on self-organizing map. IEEE Trans Inf Forensics Secur 8(4):646–656
5. Bao Z, Cao Y, Ding L, Han Z, Wang G (2008) Dynamics of load entropy during cascading failure propagation in scale-free networks. Phys Lett A 372(36):5778–5782
6. Chen J, Thorp JS, Dobson I (2005) Cascading dynamics and mitigation assessment in power system disturbances via a hidden failure model. Int J Electr Power Energy Syst 27(4):318–326
7. Buldyrev SV, Parshani R, Paul G, Stanley HE, Havlin S (2010) Catastrophic cascade of failures in interdependent networks. Nature 464(7291):1025–1028
8. Parandehgheibi M, Modiano E (2013) Robustness of interdependent networks: the case of communication networks and the power grid. In: IEEE global communications conference, pp 2164–2169
9. Vespignani A (2010) Complex networks the fragility of interdependency. Nature 464(7291):984–985
10. Bompard E, Wu D, Xue F (2011) Structural vulnerability of power systems: a topological approach 81(7):1334–1340
11. Sanchez J, Caire R, Hadjsaid N (2013) ICT and power distribution modeling using complex networks. IEEE Grenoble powertech conference, pp 1–6
12. Schneider K, Liu C-C, Paul J-P (2006) Assessment of interactions between power and telecommunications infrastructures. IEEE Trans Power Systems 21(3):1123–1130

13. Laprie JC, Kanoun K, Kaâniche M (2007) Modelling interdependencies between the electricity and information infrastructures. Comput Saf Reliab Secur 4680:54–67
14. Srivastava A, Morris TH, Ernster T, Vellaithurai C, Pan S, Adhikari U (2013) Modeling cyber-physical vulnerability of the smart grid with incomplete information. IEEE Trans Smart Grid 4(1):235–244
15. Utne IB, Hokstad P, Vatn J (2011) A method for risk modeling of interdependencies in critical infrastructures. Reliab Eng Syst Saf 96(6):671–678
16. Deng Y, Lin H, Shukla S, Thorp J, Mili L (2013) Co-simulating power systems and communication network for accurate modeling and simulation of PMU based wide area measurement systems using a global event scheduling technique. Modeling and simulation of cyber-physical energy systems workshop, pp 1–6
17. Lin H, Deng Y, Shukla S, Thorp J, Mili L (2012) Cyber security impacts on all-PMU state estimator-a case study on co-simulation platform GECO. In: Smart grid communications conference, pp 587–592
18. Li G, Ju W, Duan X, Shi D (2012) Transmission characteristics analysis of the electric power dispatching data network. Proc CSEE 32(22):141–148 (in Chinese)
19. Hu J, Li ZH, Duan XZ (2009) Structural feature analysis of the electric power dispatching data network. CSEE J Power Energy Syst 29(4):53–59
20. Li Y, Li W, Tan Y, Liu F, Cao Y, Lee KY (2017) Hierarchical decomposition for betweenness centrality measure of complex networks. Sci Rep 7:46491
21. Li C, Liu W, Cao Y, Chen H, Fang B, Zhang W, Shi H (2014) Method for evaluating the importance of power grid nodes based on PageRank algorithm. IET Gener Transm Distrib 8(11):1843–1847
22. Erdõs P, Rényi A (1960) On the evolution of random graphs. Magyar Tud Akad Mat Kutató Int Közl 5(1):17–61
23. Watts DJ, Strogatz SH (1998) Collective dynamics of 'small world' networks. Nature 393(4):440–442
24. Barabási AL, Albert R (1999) Emergence of scaling in random networks. Science 286(5439):509–512
25. Wei D, Lu Y, Jafari M, Skare PM, Rohde K (2011) Protecting smart grid automation systems against cyberattacks. IEEE Trans Smart Grid 2(4):782–795
26. Eusgeld I, Nan C, Dietz S (2011) System-of-systems approach for interdependent critical infrastructures. Reliab Eng Syst Saf 96(6):679–686
27. Parshani R, Rozenblat C, Ietri D, Ducruet C, Havlin S (2010) Inter-similarity between coupled networks. Europhys Lett 92(6):68002
28. Echenique P, Gómez-Gardeñes J, MorenoY (2004) Improved routing strategies for Internet traffic delivery. Phy Rev E 70(5):056105
29. Cao Y, Zhang Y, Bao Z (2013) Analysis of cascading failures under interactions between power grid and communication network. Elec Power Automat Equip 33(1):7–11 (in Chinese)
30. Hai Y, Wei X, Fen W (1999) The improvement of dynamic power flow calculation in dispatcher training simulator system. Automat Elec Power Sys 23(23):20–22
31. Hazra J, Sinha A (2009) Identification of catastrophic failures in power system using pattern recognition and fuzzy estimation. IEEE Trans Power Syst 24(1):378–387
32. Han P, Zhang S (2011) Analysis of cascading failures in small-world power grid. Int J Energy Sci 1(2):99–104

Chapter 3
Cascading Failure Analysis of Cyber-Physical Power System with Multiple Interdependency and Control Threshold

In this chapter, the modern infrastructure in power system is undergoing a migration to the smart grid, in which the communication network and power grid will be integrated into CPEPS. Although the traditional topological analysis revealed the mechanism of cascading failure between two networks, it ignored the control redundancy and standby lines from communication network to power grid. The robustness analysis in CPEPS requires a more comprehensive model to analyze failure behavior in reality. Here, we propose a cascading failure model with one-to-multiple interdependency and a relevant theoretical framework to analyze CPEPS cascading failure. In consideration of real CPEPS, in the proposed model, we introduce two robustness factors, the number of dependent links and control threshold, which can better describe the control function from communication nodes to power nodes. By using this model, we simulate the remaining nodes under different initial attacking nodes on high-voltage transmission network, small-world network, and double star network and analyze them under different topological combination. The simulation results show that the proposed model and robustness factors can better reveal the robustness and the mechanism of two networks in cascading failure. The present model and theoretical framework may hold a larger degree of universality, being applicable to other interdependent networks.

3.1 Introduction

With the development of communication and control technology, the traditional power grid is undergoing a change from a single power grid to an extremely large and complex multi-network coupled system which composed of the traditional power grid and communication network [1–3]. As a consequence, an initial disturbance or failure in power grid or communication network could trigger cascade failure, such as the 2003 North American blackout [4], the Italy blackout [5], and the 2004 Rome blackout [6]. Therefore, it is of much significance to analyze the mechanism of cascading failure in the cyber-physical system.

Y. Cao et al., *Cyber-Physical Energy and Power Systems*,
https://doi.org/10.1007/978-981-15-0062-6_3

Traditional power flow methods are expanded to analyze the physical side of intra-cascading failure. However, power flow calculation is more adopted in the preconceived power system accident analysis under $N-1$ contingencies [7]. When it is carried out in the fast circulating state, the complexity is more difficult to handle. Although DC power flow methods are powerful for its balance between model complexity and system behavior approximation, compared to the AC flow models [8–12], the computing scenes number of cascading fault analysis still increases exponentially with the increase of nodes number. There are some models based on the whole system characteristic, such as the optimal power flow (OPF) model [13], CASCADE model [14], influence graph model [15, 16], dynamic node model [17], dynamic load model [18, 19], and dynamical cascading failure model [20] in complex network. These methods have been studied intensively for some years, but research still focuses on the single, non-interacting power grid. Whether these methods could be extended to the CPEPS is still studied rarely.

The model of interdependent networks based on the complex networks theory develops a view of understanding cascading failures between the interdependent networks. Studies based on percolation theory show that the cascading failure transition in the one-to-one interdependent networks is first-order phenomena, while in isolated power networks, the cascading failure transition is second-order phenomena [21–23]. Then, different interface strategies such as random interface strategy, degree-to-betweenness interface strategy, and topological centrality interface strategy are simulated. It shows that the more similar the two networks are, the more robust the network is to cascading failures [24]. The previous studies mostly are based on the one-to-one interdependent networks, that is, the number of nodes in the power network is equal to the cyber network, and one power node is only dependent on one cyber node. A theoretical framework for understanding the robustness of interdependent networks with a random number of support and dependence relationship was provided [25], which extend previous works on coupled networks from one-to-one support-dependence relation to multiple support-dependence relations. Reference [26] has been observed that the interdependency between power grid and communication network is one-to-multiple. However, even after considering the multi-correspondence relationship, the coupling between the power system and the communication network is more complicated than the coupling between the two simple topologies. One important reason is that standby control lines exist in cyber-physical power system, and the communication node controls power node with some redundancy. There is little research on the relationship between this redundancy and system robustness. In this chapter, we consider the more practical control situation. The cascading failure characteristic of whole interdependent system between actual power system and double star communication network is analyzed using interdependency theory and percolation theory. In the view of the fact that control standby lines exist, each node is considered to have control margin, they need the support of a minimum number of control supply nodes to remain function. For complex network structure of power system, the coupling relationship between communication network and power grid affects the robustness of whole system. Specially, the relationship between critical point and interdependent links is analyzed. On the other hand, the robustness is also influenced by the varying of control threshold. Therefore, a mathematical model of cascading failure considering two different situations: the varying of control threshold and the varying of interdependent number are analyzed, respectively.

3.2 Modeling of the Cascading Failure in CPEPS with Control Threshold

3.2.1 Cascading Failure with One-to-Multiple Interdependency

With the development of distributed energy and the opening of electrical power market, the future smart grid requires faster control speed and better ability of big data [27, 28]. One system called quality of service is designed and aiming to better efficiency, resiliency, and flexibility than Supervisory Control and Data Acquisition system (SCADA) [1]. Besides Remote Terminal Unit (RTU) and Programmable Logic Circuits (PLC), Phasor Measurement Units (PMU) and Intelligent Electrical Device (IED) are also configured to increase the utilization of distributed and localized computations. They connect management system with data cables. The explicit modules are shown in Fig. 3.1. In the view of complex networks, the nodes coupling is the simplicity of realistic system.

There exist two types of links in interdependent networks, connectivity link and interdependent link. Connectivity link represents the intra-interdependency of each network. The function of the nodes in both networks is maintained by connectivity link w [29, 30]. For the CPEPS, connectivity link represents the transmission line in the power grid or the communication line in the cyber network [31], and the interdependent link realizes the exchange of energy or information between power grid and cyber network.

Generally, both power grid and cyber network can be expressed as unweighted undirected graphs G and C, where G represents power grid and C represents communication network. $G = (U_{\mathrm{G}}, E_{\mathrm{G}})$, $U = \{u_1, u_2, \ldots u_{\mathrm{NG}}\}$ and $C = (V_{\mathrm{C}}, E_{\mathrm{C}})$, $V = \{v_1, v_2, \ldots v_{\mathrm{NC}}\}$ are the sets of each intra-network description, respectively, $E = \{e_{ij}\}$ is the set of network connectivity links. In addition, the interdependent effects of power stations and information stations are established as set E_{I}, $E_{\mathrm{I}} = \{E_{C-G}, E_{G-C}\}$, where E_{C-G} expresses the matrix of interdependent links that cyber layer depends on the power grid. It represents cyber node u has failure without energy supply from power node v when $n_{C-P}(u, v) = 0$, where $n_{C-P}(u, v)$ is the number of interdependent links from node v in power grid to node u in communication network. The whole CPEPS is expressed as set $\xi(G, C, E, E_{\mathrm{I}})$.

Most of the research results of interdependent networks are based on the coupling ruler of one-to-one correspondence, as shown in Fig. 3.2a. While in the realistic situation, there are multiple dependencies between electric power stations and

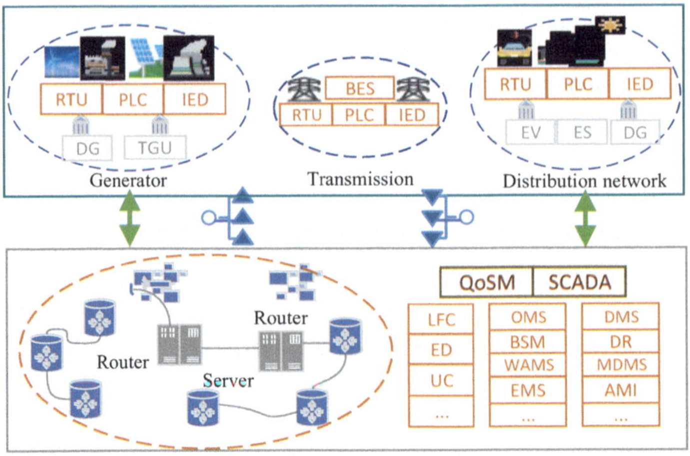

EV: Electric Vehicles
ED: Economic Dispatch
DG: Distributed Generation
LFC: Load Frequency Control
BSM: Bulk Storage Management
TGU: Traditional Generating Unit
IED: Intelligent Electronic Devices
EMS: Energy Management System
OMS: Outage Management System
ES: Energy Storage
UC: Unit Commitment
BES: Bulk Energy Storage
RTU: Remote Terminal Units
PLC: Programmable Logic Circuits
AMI: Advanced Meter Infrastructure
WAMS: Wide-Area Monitoring System
DMS: Distributed Management Systems
MDMS: Meter Data Management Systems

Fig. 3.1 Illustration on control redundancy of CPEPS

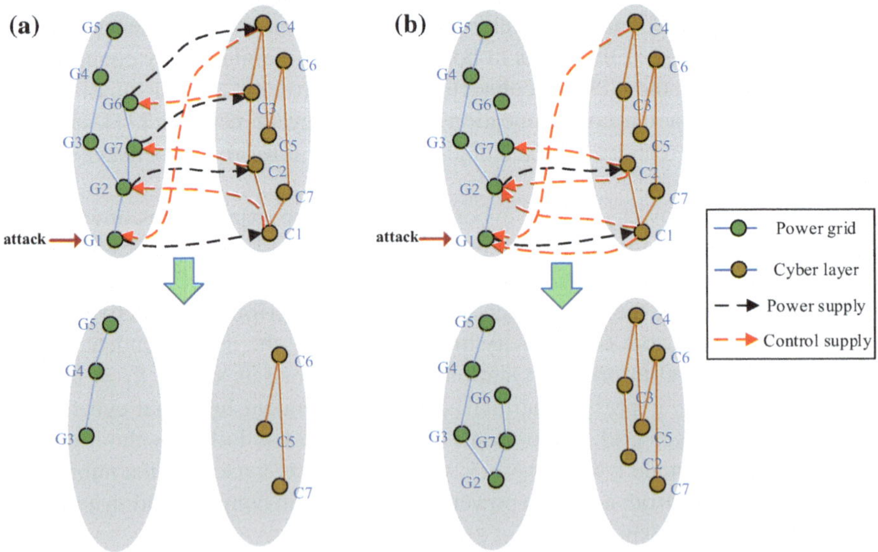

Fig. 3.2 Cyber-physical network of **a** one control interdependency and **b** multiple control interdependency with initial attack

communication stations. An electric power station can provide power for multiple information stations, and at the same time, an information station controls multiple power stations [32].

In the proposed model, each power node supplies the energy for each communication node, while each communication node controls multiple power node. The propagating process of initial attack is different between one control interdependency (Fig. 3.2a) and multiple control interdependency. The green nodes and brown nodes represent power nodes and communication nodes, respectively. In one-to-one model (Fig. 3.2a), the power supply link goes from one green node and to one brown, controlling supply link is opposite with red dotted lines. From Fig. 3.2b, one communication node has two controlling supply links, which control two different power nodes. The power supply links have not changed. Hence, the number of nodes in cyber layer is equal to that in power grid. And the number of power supply-demand interdependent links is equal to the number of nodes in power grid. The number of controlling supply-demand links is two times larger than that in one-to-one model. In general, communication node will fail when it is not connected with the power node and the same as power node. The cascading failure with initial attack in Fig. 3.2a propagates in the process of G1–C1–G2–C2–G7–C3–G6–C4, ending up in C4 because of the C4 control power node G1. While in Fig. 3.2b, system has some different from C1 to G2. Although supply control from C2 to G2 is fault, the standby link C2–G2 functionally works and supplies the necessary control. The number of interdependent links is no longer only 1. Hence, the failure nodes are just G1 and C1.

3.2.2 Cascading Failure with Control Threshold

Applying the one-to-multiple model to CPEPS, the control threshold is considered. Industrial protocol such as IEEE 61850 and DNP3 have been proposed to standardize communication between control manage center and substation. New standard of PMU and PLC will also be established. The link cables are redundant because some of multiple modular achieve same function. The corresponding relationship of cyber layer and power grid is not must be so strict that interdependent links can not break. SCADA system usually has standby communication lines for emergency control. Generally, the number of interdependent links from cyber layer to power grid has a certain redundancy in order to ensure the reliability of power supply system.

Both power grid and communication network are considered with N nodes, respectively, in real system. The initial failure nodes are $1 - p$ fraction nodes, $(1 - p)N$. Each node i in power grid has c_i control supply nodes from cyber layer. This node of power grid remains its function if the number of its functional supply nodes in the cyber layer remains greater or equal to its supply threshold $c_i^* \leq c_i$. The supply threshold is predefined as assumed for each node in power grid. In some case, some nodes in power grid are controlled by multiple communication nodes such as multiple wide-area control. The control threshold represents the power nodes minimal necessary interdependent links from cyber layer.

We study two conditions should be satisfied if a node in power grid works functionally above.

(1) The node belongs to the giant component in its own network.
(2) At lease c_i^* control supply interdependent links of this node exist, where these links come from other functional nodes in the cyber layer.

So the cascading failure model first build is that firstly a random fraction $1 - p$ of the nodes in power grid are attacked, then we calculate the giant component of power grid, then the failure judgment transfers to cyber layer. The cascading failure is divided into several steps. In the k step, we firstly judge the power grid

$$n_{C-G(i)}^{k} < C_i^* \tag{3.1}$$

where $n_{G-C(i)}^{k}$ is the number of real-time interdependent links of node i in power grid supplied from cyber layer in failure process. That is, one power station is easier to failure than they used to be, since it needs control supply from more information stations. While the communication nodes in this chapter are set to be supplied by single power station. Then the cyber layer is judged by

$$n_{C-G(i)}^{k} < 1 \tag{3.2}$$

Each node i in cyber layer has $n_{C-G}^{k}(i)$ power supply nodes in the power grid that connected to node i by supply links.

3.3 Robustness Evaluation of CPEPS in Cascading Failure

The characteristic of cascading failure in two coupling networks is different from the second-order phase transition in single intra-network. It is first-order phase transition [21]. At the ending of cascading failure, the probability of randomly selected node belonging to the giant component is μ.

We analyze the dynamic of cascading failure using percolation theory in this chapter. The power grid G and cyber layer C have the degree distribution $P_{\mathrm{G}}(k)$ and $P_{\mathrm{C}}(k)$, when $1 - p$ of power nodes are randomly attacked, the exacerbation factor of power grid is

$$g_{\mathrm{G}}(p) = 1 - G_0^G[1 - p(1 - f_\infty^G)] \tag{3.3}$$

where f_∞^G is the probability that satisfied the transcendental equation

$$f_\infty^G = G_1^G[1 - p(1 - f_\infty^G)] \tag{3.4}$$

And that of communication network have the same form. The generating function of the degree distribution is $G_0^G(x) = \sum_k P_G(k)x^k$. Analogous, the generating function of excess degree distribution is $G_1^G(x) = G_0^{G'}(x)/G_0^{G'}$ [33, 34].

For simplicity, we assume that function $p_s(j, c)$ is a probability that a node with c supply links works functionally if j of its c supply nodes in the cyber layer work functionally, then a cumulative probability distribution of power grid is calculated

$$t_{sG}(j, c) = P(c_{sG}^* \le j | c_{sG} = c) \tag{3.5}$$

where $P(\cdot)$ is the conditional probability. The cumulative probability distribution of communication network is the same. From conventional homogeneous k-core percolation [35–39], we introduce the function $H_{sG}(x)$, $H_{sC}(x)$, $L_{sG}(x)$, $L_{sC}(x)$.

$$H_{sG}(\gamma) = \sum_{c=0}^{\infty} P_{sG}(c) \sum_{j=0}^{c} \binom{c}{j} t_{sG}(j, c)\gamma^j (1-\gamma)^{c-j} \tag{3.6}$$

$$L_{sG}(\gamma) = \sum_{c=0}^{\infty} \frac{cP_{sG}(k)}{< c_s >} \sum_{j=0}^{c-1} t_{sG}(j+1, c)\gamma^j (1-\gamma)^{c-j} \tag{3.7}$$

where $H_{sG}(x)$ and $L_{sG}(x)$ are the k-core generating functions of degree distribution and excess degree distribution of supply links in power grid. The cumulative distribution of threshold in power grid is finally simplified to

$$t_{sG}(j, c) = \begin{cases} 0, & c_{sG}^* < j \\ 1, & c_{sG}^* \ge j \end{cases} \tag{3.8}$$

And that in cyber layer are

$$H_{sC}(\gamma) = \sum_{c=0}^{\infty} P_{sC}(c) \sum_{j=0}^{c} \binom{C}{j} t_{sC}(j, c)\gamma^j (1-\gamma)^{c-j} \tag{3.9}$$

$$L_{sC}(\gamma) = \sum_{C=0}^{\infty} \frac{cP_{sC}(c)}{< c_s >} \sum_{j=0}^{c-1} t_{sC}(j+1, c)\gamma^j (1-\gamma)^{c-j} \tag{3.10}$$

The cumulative distribution of threshold in communication network is given by

$$t_{sG}(j, c) = \begin{cases} 0, & j < 1 \\ 1, & j \ge 1 \end{cases} \tag{3.11}$$

We analyze the percolation step of failure propagation between power grid and communication network. In stage 1, after remove $1 - p$ fraction of nodes, the surviving fraction of power grid is determined by condition (1). It can be expressed in the closed-form expression

$$\mu_{G.1} = pg_G(p) \tag{3.12}$$

The probability of that random interdependent links is corresponded to functionally work nodes in power grid is

$$f_{C,1} = \mu_{G,1} \tag{3.13}$$

In stage 1, it equals to the fraction of surviving power nodes. Then, the remaining fraction of communication network is considered. The initial failure nodes in communication network caused by broke interdependent links is

$$y_{C,1} = H_{sC}(f_{C,1}) \tag{3.14}$$

The surviving fraction of communication network is

$$\mu_{C,1} = y_{C,1}g_C(y_{C,1}) \tag{3.15}$$

However, when the failure is repropagated from communication network to power grid, it no longer simply follows the probability relationship of the probability $\mu_{C,1}$, so $f_{G,2} \neq \mu_{C,1}$. Because of the control margin, the situation that all the power nodes corresponding to the failed control links can not work no longer exists.

Therefore, we must first analyze the failure probability of interdependent links from communication network to power grid in stage 2. If one interdependent link corresponds to node i in power grid and node j in communication network, then the probability of its surviving depends on how many other links (belong to node i or j) survive. Then, $f_{G,2}$ should multiply the generating function of excess degree distribution L_{sC}.

$$f_{G,2} = L_{sC}(f_{C,1})g_C(y_{c,1}) \tag{3.16}$$

The $f_{G,2}$ is calculated and then the probability of unfunctionally working power nodes caused by the out-of-order interdependent links can be calculated, just using generating function of degree distribution $H_{sG}(\cdot)$, which is similar to stage 1.

$$y_{G,2} = pH_{sG}(f_{G,2}) \tag{3.17}$$

Then, the size of finite components in power grid caused by its intra-dependency is

$$\mu_{G,2} = y_{G,2}g_G(y_{G,2}) \tag{3.18}$$

In same way, the recursion relations for the stages $n > 1$ are

$$f_{G,n} = L_{sC}(f_{C,n-1})g_C(y_{C,n-1}) \tag{3.19}$$

$$f_{C,n} = pL_{sG}(f_{G,n})g_G(y_{G,n}) \tag{3.20}$$

where

$$y_{G,n} = pH_{sG}(f_{G,n}) \tag{3.21}$$

$$y_{C,n} = H_{sC}(f_{C,n}) \tag{3.22}$$

The fractions of functional nodes at stage n in the cascade failure are

$$\mu_{G,n} = y_{G,n}g_G(y_{G,n}) \tag{3.23}$$

$$\mu_{C,n} = y_{C,n}g_C(y_{C,n}) \tag{3.24}$$

We have the obtained equation in terms of f_G at the steady state

$$f_{G,n} = f_{G,n-1} = f_G \tag{3.25}$$

$$f_G = L_{sC}(f_C)[H_{sC}(f_C)] \tag{3.26}$$

where

$$f_C = pH_{sG}(f_G)g_G[pH_{sG}(f_G)] \tag{3.27}$$

Graph solution is used to find the critical point p_c, which should satisfy

$$\frac{\mathrm{d}F(f_A)}{\mathrm{d}f_A} = 1 \tag{3.28}$$

The Eqs. (3.26) and (3.28) can be used to calculate the critical point p_c.

The mechanism of cascading failure in Sect. 3.2 can be analyzed through above formula. We measured the relationship between probability μ and initial remaining ratio p of nodes after attacked, shown as $p-\mu$ curve. Although the size of smart grid is not as large as complex network, this curve is still effective in the aspect of cascading failure. In the curve, μ changes obviously. When $p < p_c$, μ is approximately 0, and $p > p_c$, $\mu > 0$. Obviously, the critical point p_c is an index of the system robustness in cascading failure. The smaller critical point is, the more initial attacking ratio needs for whole system collapse [35].

3.4 Case Studies

Our tested power grid is taken from the high-voltage transmission system in Hunan Province, China, which has 241 nodes and edges. The details are described in Table 3.1. This network is one typical small-world network. It satisfied

$$C \gg C_{\text{random}} \tag{3.29}$$

$$L \geq L_{\text{random}} \tag{3.30}$$

where C_{random} is the *Clustering Coefficient* in random network with the same nodes of the HV transmission system. L_{random} is the *Characteristic Path Length* in random network with the same nodes of the HV transmission system [40, 41]. For a general situation to reflect power grid, we used the small world with 2000 nodes to simulation. The details are shown as Table 3.1. N and M are the numbers of nodes and edges. <*k*> is the average degree of the network.

To compare the influence of communication network topology on the cascading failure of CPEPS, two topologies of communication network are simulated. The SW network in communication network has the same topology as the power grid [31]. While the double star (DS) communication networks are the scale-free networks, whose degree distribution follows a power law, $P_C(k) \propto k^{-\gamma}$, where $P(k)$ is the probability that the degree of a node is k, γ is power law exponent. The details are shown as Table 3.2.

When a new failure occurs in the power grid, the intra-dependency would be disrupted, and some power grid nodes have to be removed. Then the failure spreads through interdependent links. The similar process occurs in communication networks. The interdependency from communication network to power grid makes the failure spread to power grid again. Specifically, the following simulating steps are performed.

Table 3.1 Topology parameters in power grid

	N	*M*	<*k*>	*C*	*L*
HV transmission system	241	371	3.08	0.19	5.87
Random network	241	365	3.08	0.0168	4.98
SW network in power grid	2000	3562	3.50	0.19	6.08

Table 3.2 Topology parameters in communication network

	N	*M*	<*k*>	*C*	*L*
SW communication network	2000	3562	3.50	0.19	6.08
DS communication network	2000	3600	3.60	0.95	1.98

Step 1: Generate random failure nodes in power grid. In each simulation, we vary the size of the initiating attacking proportion, $1 - p$, which the number of power nodes in the initial random failure is $(1 - p) \times N$.
Step 2: Calculate the remaining nodes in power grid. Because the nodes fault causes related links break, the intra-dependencies work unfunctionally, and the failure in power grid will spread. The remaining nodes after the intra-failure are calculated according to intra-dependent links.
Step 3: Judge whether the number of remaining nodes in power grid is 0 or it is equal to the number in last *Step* 2. If true, the cascading failure is ending. Else turn to *Step* 4.
Step 4: Mark failure nodes in communication network for interdependent links fault and remove failure nodes in power grid. According to the supply relationship from power grid to communication network, if $n_{C-G(i)} < 1$, then the node i in communication network has fault. After the failure nodes are marked, remove the failure nodes of power grid in *Step* 2.
Step 5: Calculate the remaining nodes in communication network. It is the same as *Step* 2.
Step 6: Judge whether the number of remaining nodes in communication network is 0, or it is equal to the number in last *Step* 4. If true, the cascading failure is ending. Else turn to *Step* 7.
Step 7: Mark failure nodes in power grid for interdependent links fault and remove failure nodes in communication network. According to the supply relationship from communication network to power grid, if $n_{C-G(i)} < c^*_{\mathrm{sG}}$, then the node i in power grid has fault. After the failure nodes are marked, remove the failure nodes of communication network in *Step* 5. Turn to *Step* 2.

3.4.1 Impacts of Different Interdependent Links

First, we analyze the p-μ curve with the number of interdependent links in SW-SW coupling networks model. The result is shown as Fig. 3.3. The system's robustness increases with the increase of the number of interdependent links. Given the number of interdependent links with 1, 2, 3, 5, and 8, they responded five p-μ curves. The critical point $p_c = 0.58$ when $n = 1$, and p_c is 0.32 when $n = 8$. On the other hand, there is saturation when the interdependent links reach to certain point. The difference of μ between $n = 5$ and $n = 8$ is negligible. The saturation number of interdependent links is 5 from the histograms of $n - p_c$ in Fig. 3.4.

In addition to reflect the influence of networks topology on cascading failure, we compare the SW-SW networks and SW-DS networks. Double star network is common in communication. When SW-SW model is simulated, both power grid and communication network have the same result as shown in Fig. 3.3, although the interdependent from power grid to communication network and that from communication network to power grid are different. When SW-DS model is simulated,

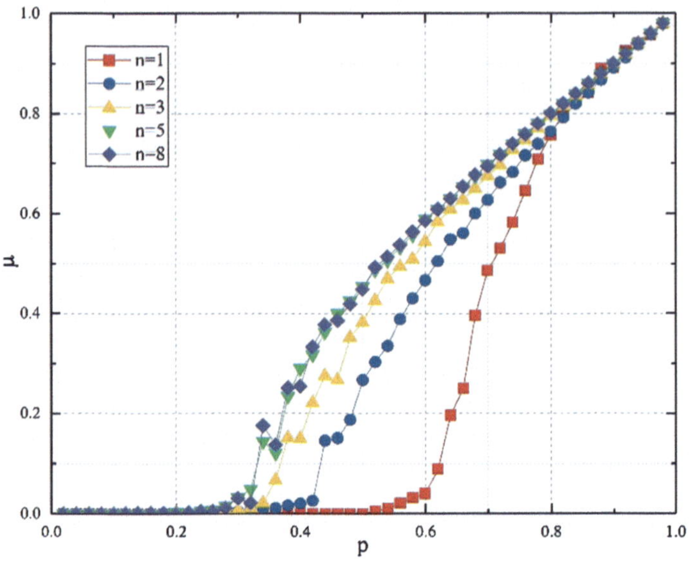

Fig. 3.3 Robustness of SW-SW coupled networks to random failures, with varying numbers of interdependent links n

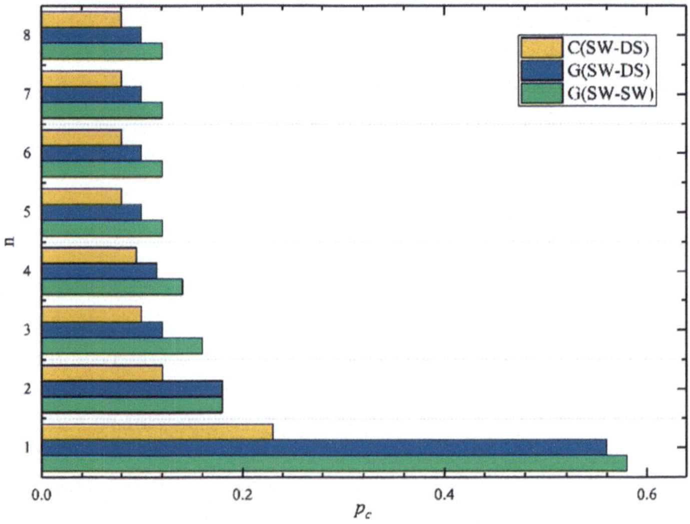

Fig. 3.4 Critical point of three models with varying numbers of interdependent links n

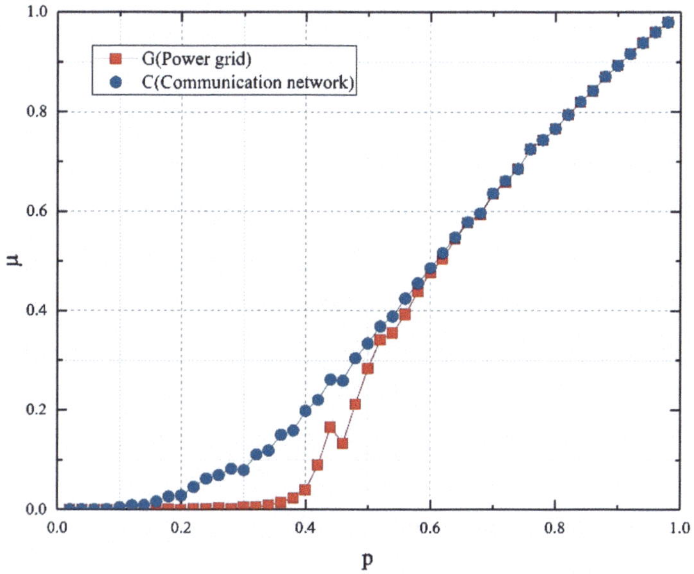

Fig. 3.5 Robustness comparison between power grid and communication network in SW-DS model with $n = 2$

Fig. 3.6a reflects the curve of power grid while Fig. 3.6b reflects that of communication network. The p-μ curve between power grid and communication network is compared as shown in Fig. 3.5, when p is in a high interval, there is no difference of the μ value changing between power grid and communication network. The topology influences the low-value interval of p. DS communication network has the lower critical point 0.20. Obviously, in this model, the communication network is more robust to random failures than power grid.

Then, the problem is which model is more resilience for power grid to random attacks. As shown in Fig. 3.4, the critical point p_c of power grid in SW-SW model is always larger than that in SW-DS model. The topology of communication network influences the critical point of whole systems in cascading failure. The double star network has more resilience for the whole system. The DS communication networks is a scale-free one and the operation centers which control power nodes and exchange information with other communication devices are some autonomous nodes. Thus, the double star structure dispatching data network for the power system is better in case of random attacks.

3.4.2 Impacts of Different Control Threshold

In addition, Fig. 3.7 reflects the results when the control threshold is considered. Similar observation as the number of interdependent links can be found, the system

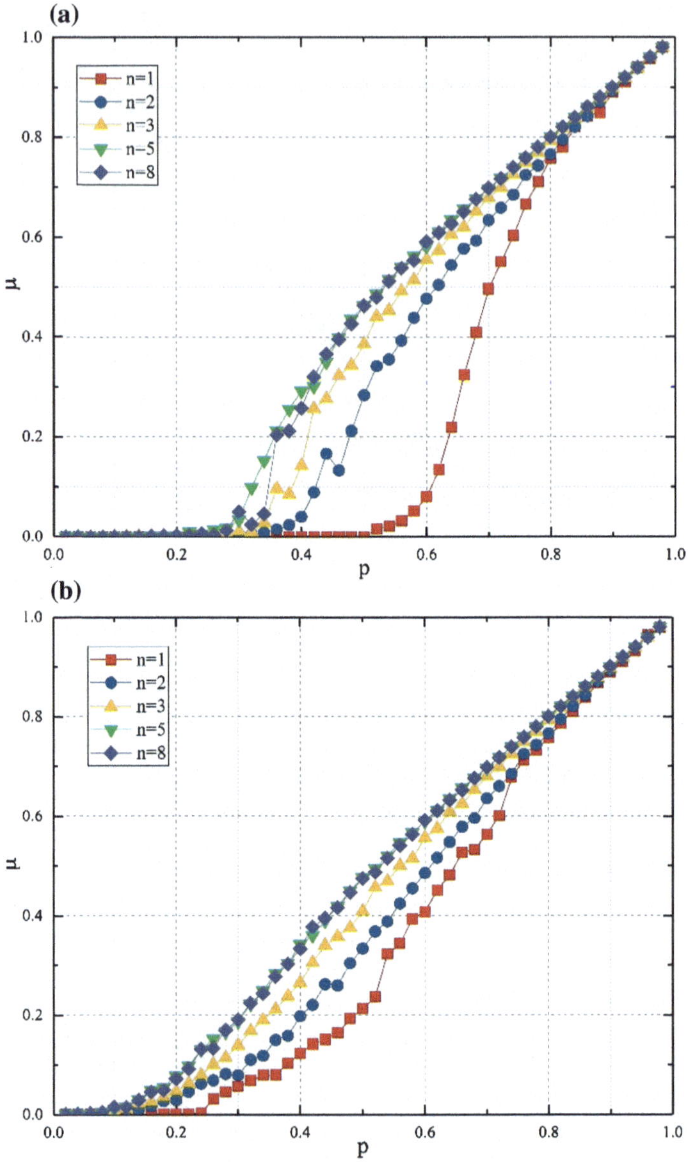

Fig. 3.6 Robustness of SW-DS coupled networks to random failures in power grid (**a**) and communication network (**b**), with varying numbers of interdependent links n

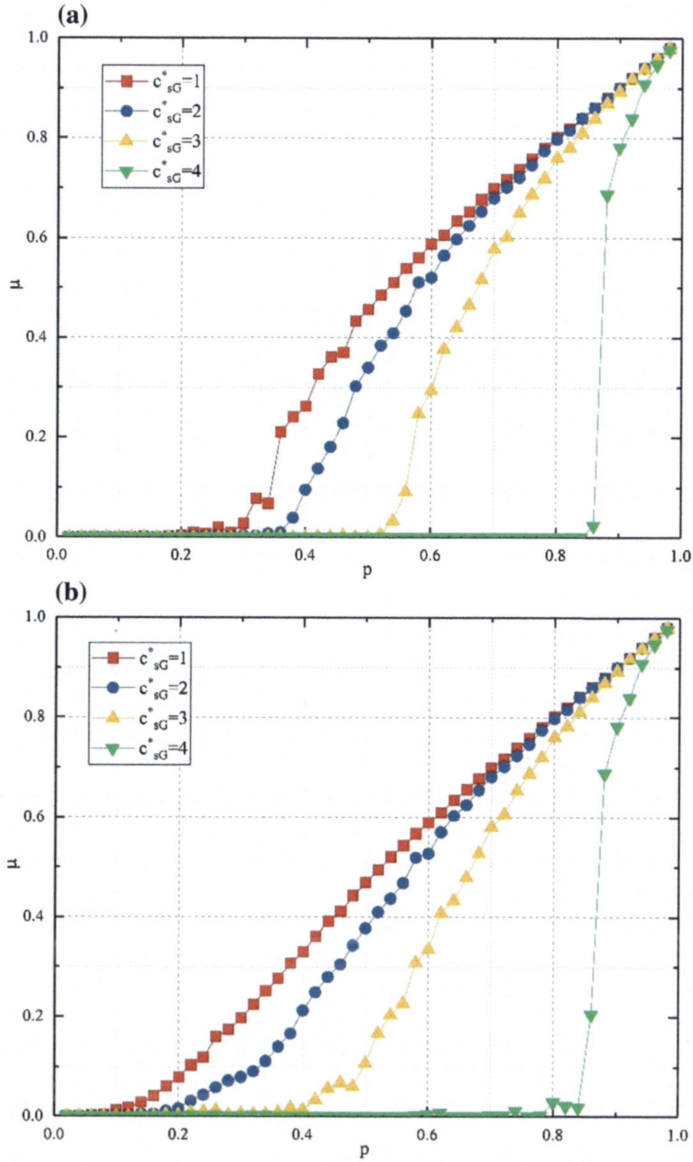

Fig. 3.7 Robustness of SW-DS coupled networks to random failures in power grid (**a**) and communication network (**b**), with varying numbers of control threshold c^*_{sG}

Table 3.3 Different robustness to same redundant values

	Redundant values in power grid			Redundant values in communication network		
Value	1	2	3	1	2	3
$c_{sG}^{*} = 1$	0.860	0.510	0.380	0.830	0.400	0.210
$n = 5$	0.180	0.120	0.115	0.12	0.100	0.095

robustness increases with the decrease of control threshold c_{sG}^{*}. However, the specific influence of the system robustness depends both on interdependency and control threshold. The control threshold subtracts from interdependent links number $n\text{-}c_{sG}^{*}$ is the redundant value of lines. As shown in Table 3.3, although redundant values are same, different control thresholds and different interdependent link numbers still make different robustness to cascading failure.

3.5 Summary

In this chapter, we introduce a model of cyber-physical system in cascading failure. By introducing the control threshold, the cascading failure model with the existence of redundancy and standby lines in control supply is developed. With the one-to-multiple interdependent relationship, the critical point based on percolation theory is measured to evaluate the robustness of CPEPS. The topology simulator of HV transmission network with 241 nodes and 371 edges is implemented in Python. For further enhancing the application of the model in any power grid, we also analyze a SW network with 2000 nodes. For communication network, SW network and DS network are compared to find the robustness impact on topology, interdependent link, and control threshold.

The simulation extends the robustness factors in cascading failure of CPEPS. The control threshold presented in this chapter can reflect the coupling strength from communication network to power grid. Both increasing the number of interdependent links and decreasing the control threshold have the saturation to enhance the robustness of CPEPS. And as a common topology in communication network, we verify the DS communication network's effectiveness in resisting cascading failure both for the intra-network and the whole system.

References

1. Tomsovic K, Bakken D, Venkatasubramanian V, Bose A (2005) Designing the next generation of real-time control, communication, and computations for large power systems. Proc IEEE 93(5):965–979

2. Amin M (2001) Toward self-healing energy infrastructure systems. IEEE Comput Appl Power 14(1):20–28
3. Nobile E, Bose A (2002) A new scheme for voltage control in a competitive ancillary service market. In: Power systems computation meeting
4. Final report on the August 14, 2003 blackout in the United States and Canada: causes and recommendations. US-Canada power system outages task force, pp 1691–1702
5. Corsi S, Sabelli C (2004) General blackout in Italy sunday september 28, 2003, h. 03:28:00. In: IEEE power engineering society general meeting
6. Bobbio A, Bonanni G, Ciancamerla E, Clemente R, Iacomini A, Minichino M, Scarlatti A, Terruggia R, Zendri E (2010) Unavailability of critical SCADA communication links interconnecting a power grid and a Telco network. Reliab Eng Syst Saf 95(12SI):1345–1357
7. Cao Y, Wang G, Han Z, Ding L, Bao Z, Cao L (2009) A cascading failures model in power grid considering topology evolvement. Autom Electron Power Syst 33(9):5–10 (in Chinese)
8. He H, Yan J (2016) Cyber-physical attacks and defences in the smart grid: a survey. IET Cyber-Phys Syst Theor Appl 1(1):13–27
9. Yan J, Tang Y, He H, Sun Y (2015) Cascading failure analysis with DC power flow model and transient stability analysis. IEEE Trans Power Syst 30(1):285–297
10. Al-Takrouri S, Savkin AV, Agelidis VG (2013) A decentralized control algorithm based on the DC power flow model for avoiding cascaded failures in power networks. In: Asian control conference (ASCC), pp 1–6
11. Cetinay H, Soltan S, Kuipers F A, Zussman G, Van Mieghem P (2017) Comparing the effects of failures in power grids under the AC and DC power flow models. IEEE Trans Netw Sci Eng 1
12. Mei S, Ni Y, Wang G, Wu S (2008) A study of self-organized criticality of power system under cascading failures based on AC-OPF with voltage stability margin. IEEE Trans Power Syst 23(4):1719–1726
13. Mei SW, Weng XF, Xue AC (2006) Blackout model based on OPF and its self-organized criticality. In: Chinese control conference, pp 1673–1678
14. Dobson I, Carreras BA, Lynch VE, Newman DE (2001) An initial model for complex dynamics in electric power system blackouts. In: Proceedings of the 34th Hawaii international conference on power system sciences, pp 710–718
15. Wei X, Zhao J, Huang T, Bompard EF (2018) A novel cascading faults graph based transmission network vulnerability assessment method. IEEE Trans Power Syst 33(3):0885–8950
16. Wei X, Gao S, Li D, Tao H, Pi R, Tao W (2018) Cascading fault graph for the analysis of transmission network vulnerability under different attacks. Proc CSEE 38(2):465–474
17. Moreno Y, Gomez J, Pacheco A (2002) Instability of scale-free networks under node-breaking avalanches. Europhys Lett 58(4):630–636
18. Bao Z, Cao Y, Ding L, Han Z, Wang G (2008) Dynamics of load entropy during cascading failure propagation in scale-free networks. Phys Lett A 372(36):5778–5782
19. Bao Z, Cao Y (2008) Cascading failures in local-world evolving networks. J Zhejiang Univ-Sci A 9(10):1336–1340
20. Ding L, Cao Y, Wang G, Liu M (2011) Dynamical model and analysis of cascading failures on the complex power grids. Kybernetes 40(5):814–823
21. Buldyrev SV, Havlin S, Parshani R, Paul G, Stanley HE, Havlin S (2010) Catastrophic cascade of failures in interdependent networks. Nature 464(7291):1025–1028
22. Zhu Y, Yan J, Sun Y, He H (2014) Revealing cascading future vulnerability in power grids using risk-graph. IEEE Trans Parallel Distrib Syst 25(12):3274–3284
23. Brummitt CD, Souza RM, Leicht EA (2012) Suppressing cascades of load in interdependent networks. Proc Natl Acad Sci USA 109(12):E680–E689
24. Parandehgheibi M, Modiano E, Hay D (2016) Mitigating cascading failures in interdependent power grids and communication networks. In: IEEE international conference on smart grid communications, pp 242–247
25. Shao J, Buldyrev SV, Havlin S, Stanley HE (2011) Cascade of failures in coupled network systems with multiple support-dependence relations. Phys Rev E 83(2):1127–1134

26. Huang Z, Wang C, Ruj S, Stojmenovic M, Nayak A (2013) Modeling cascading failures in smart power grid using interdependent complex networks and percolation theory. In: IEEE 8th conference on industrial electronics and applications, pp 1023–1028
27. Cai Y, Cao Y, Li Y, Huang T, Zhou B (2016) Cascading failure analysis considering interaction between power grids and communication networks. IEEE Trans Smart Grid 7(1):530–538
28. Macana CA, Quijano N, Mojica-Nava E (2011) A survey on cyber physical energy systems and their applications on smart grids. In: IEEE PES conference on ISGT LA, pp 1–7
29. Bose A (2003) Power system stability: new opportunities for control. In: Stability and control of dynamical systems with applications
30. Sridhar S, Hahn A, Govindarasu M (2011) Cyber–physical system security for the electric power grid. Proc IEEE 100(1):210–224
31. Zhang Y, Cao Y, Bao Z (2012) Impact of transmission distortion of line-outage-state information on cascading failures. Automat Electron Power Sys 36(24):4–9 (in Chinese)
32. Palensky P, Widl E, Elsheikh A (2014) Simulating cyber-physical energy systems: challenges, tools and methods. IEEE Trans Syst Man Cybern Syst 44(3):318–326
33. Carreras BA, Newman DE, Dobson I, Poole AB (2004) Evidence for self-organized criticality in a time series of electric power system blackouts. IEEE Trans Circ Syst. I: Regul Pap 51(9):1733–1740
34. Carreras BA, Newman DE, Dobson I, Poole AB (2000) Initial evidence for self-organized criticality in electric power system blackouts. In: Proceedings of the 33rd annual Hawaii international conference on system sciences, pp 1–6
35. Cellai D, Lawlor A, Dawson KA, Gleeson JP (2013) Critical phenomena in heterogeneous k-core percolation. Phys Rev E 87(2):022134
36. Bao Z, Cao Y, Ding L, Wang G (2009) Comparison of cascading failures in small-world and scale-free networks subject to vertex and edge attacks. Phys A Stat Mach Appl 388(20):4491–4498
37. Carmi S, Havlin S, Kirkpatrick S, Shavitt Y, Shir E (2007) From the cover: a model of internet topology using k-shell decomposition. Proc Natl Acad Sci 104(27):11150–11154
38. Dorogovtsev SN, Goltsev AV, Mendes JFF (2006) k-core organization of complex networks. Phys Rev Lett 96(4):185–194
39. Goltsev AV, Dorogovtsev SN, Mendes JFF (2006) k-core(boot-strap) percolation on complex networks: critical phenomena and nonlocal effects. Phys Rev E 73:056101
40. Xu L, Wang X, Wang X (2009) Equivalent admittance small-world model for power system-I. Basic concepts and implementation. In: 2009 Asia-Pacific power and energy engineering conference, pp 1–4
41. Ding M, Han P (2006) Reliability assessment to large-scale power grid based on small-world topological model. In: 2006 International conference on power system technology, pp 1–5

Chapter 4
Impacts of EPON-Based Communication Networks on Differential Protection of Smart Distribution Networks

Problems in smart distribution networks, such as multidirectional flow, rapid change of load, and three-phase unbalance, bring a big challenge on the reliability of the power grid. In order to solve these problems, various protection methods have been proposed. The differential protection is being applied in micro-grids and distribution networks for its excellent performance. In literature [1], the current differential protection is used in the loop of the future renewable electric energy delivery and management (FREEDM) system, and both the sensitivity and the rapidity are verified. Similarly, a new protection method using the differential current and the restraint current is proposed to identify the fault [2]. To eliminate the impact of inverter-interfaced distributed generation (IIDG) on protection, the negative-sequence current is used as the auxiliary criterion for the current differential protection [3]. In literature [4], multiple diff rings are proposed to detect and locate the fault in ship, which can distinguish the physical and the sensor fault. Unlike the transmission grid, the distribution network has various power quality problems, such as harmonic and imbalance, and these problems inevitably affect the current differential protection performance. To address these problems, some improved solutions are proposed in [5, 6].

For best performance and fastest clearing fault times, the differential protection algorithm is often performed at the intelligent terminal unit (ITU), which needs to know the data of other ITUs at any given time. Therefore, the point-to-point communication architecture is used widely. In literature [2], the fiber-optical Ethernet is used for the communication between each ITU. In literature [7], the Zigbee technology has been applied for the Directional Comparison Pilot Protection to reduce the transmission channel cost. In literature [8], the wireless token ring protocol is further used to achieve a better data sharing. However, the contradiction between high investment cost and high performance inhibits its application in distribution networks.

Ethernet Passive Optical Network is a passive optical network communication technology, which has the advantages of high speed, convenient maintenance, and low price. It has been widely applied in the communication system of smart distribution networks for monitoring and control [9]. In literature [10], an access control

Y. Cao et al., *Cyber-Physical Energy and Power Systems*,
https://doi.org/10.1007/978-981-15-0062-6_4

mechanism for distribution automation with a communication network using Ethernet Passive Optical Network (EPON) is described. A protection scheme of EPON-based communication network for offshore wind power farms has been proposed in [11]. An adaptive admission control algorithm is introduced in [12], which can provide quality of service (QoS) support for the Über-FiWi network combining with EPON and WiMAX. In literature [13], the reliability of communication system in smart distribution grid based on EPON is analyzed, and a topology with the best total reliability of communication is proposed.

If differential protection can be applied to EPON-based communication network, not only the distribution network can get better protection, but also the investment cost can be greatly reduced. However, EPON is a point to multi-point network, and ITUs cannot communicate with each other directly. The calculation process and the impact of communication network on the protection are quite different from the traditional network. Therefore, the study of these topics is important for the application of differential protection. This chapter focuses on the impacts of EPON-based communication network on the differential protection and analyzes the feasibility of the application.

4.1 Overview of Differential Protection Algorithms

4.1.1 Principle of Current Differential Protection (CDP)

For a protection zone, according to the Kirchhoff's current law (KCL), one can obtain the deduction that the summation of the current vector at each node in this zone should be equal to zero, and if not, there must be a fault in it. The same is true for the protected device. Considering a distribution network shown in Fig. 4.1, this network can be divided into several zones, and the summation can be calculated for each zone. Once the short-circuit fault happens, the fault zone like Z_4 can be located and isolated accordingly, and the non-faulty zones are not affected.

The expression of the summation I_{sum} is

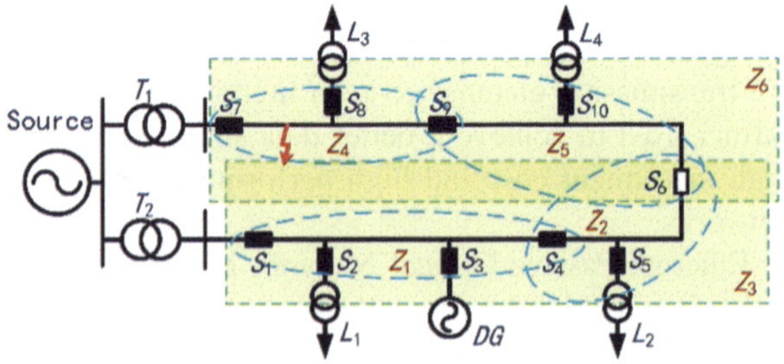

Fig. 4.1 Division of the feeder by using the current differential protection

$$I_{sum}(t) = \left| \sum_{k=1}^{N} I_k(t) \right| \quad (4.1)$$

where I_k is the current vector or the positive sequence current of the node k; N is the total number of nodes.

With the existence of the measurement error, the I_{sum} is not equal to zero, and a minimum threshold is considered to limit this bias. The criterion of fault occurrence is:

$$I_{sum}(t) \geq I_0 \quad (4.2)$$

where I_0 is the threshold of I_{sum}.

Since the value of I_0 is set according to experience and not easy to be used, another criterion is proposed as follows

$$I_{sum}(t) \geq S_0 \sum_{k=1}^{N} |I_k(t)| \quad (4.3)$$

where S_0 is a restraint coefficient. In (4.3), the threshold is calculated by I_k and can be adjusted according to the environment. But the setting of the coefficient still needs experience.

The CDP has many advantages. At first, the CDP can quickly remove the fault in the whole protection area. Since the CDP operates only when the fault occurs in the protection area, it does not have the problem of selective coordination with the adjacent protection elements. At the same time, even if the system has oscillations or incomplete phase operation, the stability of CDP will not be affected since the judgment mechanism does not change. However, the CDP needs the high-performance communication channel to ensure the synchronization of all current vectors in protected zone. Once the information is out of synchronization, the CDP will fail, which will lead to missing or misjudged.

4.1.2 Principle of Directional Comparison Pilot Protection

Since the CDP needs the high-performance communication channel, it is often replaced by the Directional Comparison Pilot Protection method. The Directional Comparison Pilot Protection (DCPP) does not require strict time synchronization or fast transmission channels and can work under the poor network conditions.

Due to the direction measuring component, the DCPP can be divided into many types, such as the Latching Directional Comparison Pilot Protection (LDCPP) and so on. But the process of most DCPPs can be divided into two phases: start phase and action phase. In the start phase, the status of the protected zone is checked, and the criterion of status is as follows

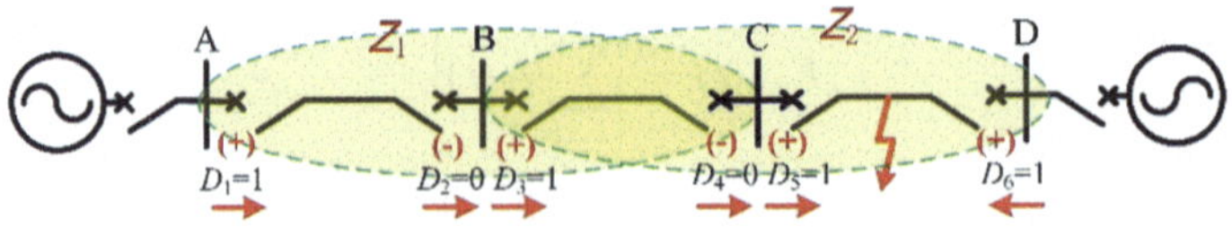

Fig. 4.2 Direction judgment of the DCPP

$$I_k(t) > I_S \quad k \in (1, 2, \ldots, N) \tag{4.4}$$

where I_s is the threshold of starting current, sometimes, the voltage can also be used for the judgment in start phase.

After detecting the abnormal status, the action phase is started to locate the fault. The criterion is as follows

$$\prod_{k=1}^{T} D_k(t) = 1 \tag{4.5}$$

where D_k is the current direction of the node k. T is the total number of overcurrent nodes. Specifying that the current flows out of the busbar is in the positive direction, D_k is equal to 1 when the overcurrent is in the positive direction and is equal to 0 when the overcurrent is in the negative direction.

As shown in Fig. 4.2, if the short-circuit fault occurs in the L_{CD}, where D_1, D_3, D_5, D_6 are the positive direction, and the D_2, D_4 are the negative direction. By Formula (4.5), the location of the fault can be determined.

As the DCPP only needs to calculate the current direction which is represented as a binary, there will be no data error caused by time delay and time synchronization error.

4.1.3 Principle of Backup Differential Protection

The differential protection consisting of adjacent switches can isolate faults in the protection zone in the shortest time. However, once the protection fails, the backup differential protection will work immediately. The backup zone is a larger area that is made up of two or more protection zones. In the backup zone, the differential protection algorithm is used to determine if the fault is in this area and is calculated in parallel with the differential protection in normal zones. Nodes involved in the backup protection are nodes connected to the outside of the backup zone. In Fig. 4.1, Z_3 is the backup zone of Z_1 and Z_2, and the switches involved in calculation are S_1–S_3, S_5 and S_6; Z_6 is the backup zone of Z_4 and Z_5, and the involved switches are S_6–S_8 and S_{10}.

4.2 Calculation Process of Differential Protection Based on EPON

4.2.1 Calculation Process

The EPON adopts point to multi-point structure and the transmission type for uplink and downlink is different. In the uplink transmission, only one optical network unit (ONU) signal can arrive at the optical line terminal (OLT) in a given period. In the downlink transmission, the OLT broadcast the data to the network, and the ONU selects their own data by the logical link identifier (LLID). Under this structure, ITUs cannot communicate with each other directly, and all data must be exchanged on the server. The calculation process is executed following the below steps:

Step 1: The current data from each ITU will be sent to the server periodically.

Step 2: The server receives data from ITUs and checks whether all data is received. If not, then wait.

Step 3: The server calculates the I_{sum} of each protected zone, determines if a fault has occurred, and returns the operation instruction to ITUs which belong to the fault zone.

Step 4: ITU receives and executes the operation instruction.

4.2.2 Long Distance Communication of EPON

Due to the limited link budget of EPON, the transmission distance of EPON is limited within 20 km [14]. However, the line length of the actual distribution network is usually larger than this distance. A feasible solution is to divide the line into multiple segments, establish an EPON for each section, and then connect multiple EPON with optical fiber. The multi-service transport platform (MSTP) based on the synchronous digital hierarchy (SDH) are usually used for the communication management [15].

4.2.3 Communication Delay

The complete communication network for smart distribution networks is combining with SDH network and EPON, as shown in Fig. 4.3.

The time delay evaluation of SDH can be expressed by [13]:

$$T_{\text{SDH}} = t_e + t_r \times n + t_o \times l + d/R \tag{4.6}$$

where t_e is the delay caused by the add-drop multiplexer (ADM), in which the Ethernet data is converted to the optical signals or the optical signals are converted to the Ethernet data, t_r is the delay for data relay, and this delay is usually caused by

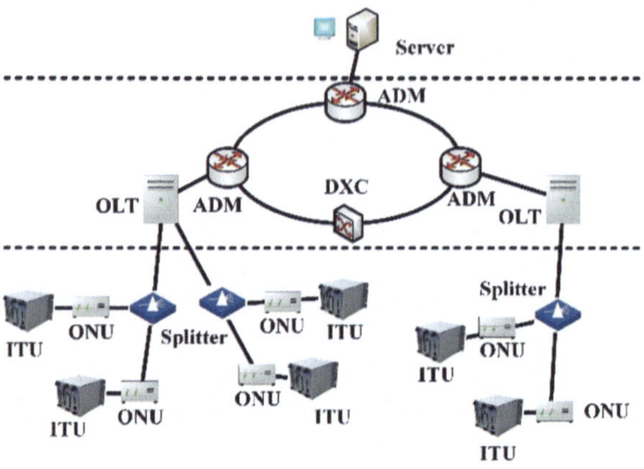

Fig. 4.3 Communication network for smart distribution networks

ADM or digital cross-connect (DCC), n is the number of relay equipment, t_o is the transmission delay of the optical signal in the optical fiber line per kilometer, l is the length of fiber optic lines, R is the communication rate of SDH, and d is the length of a particular generic object oriented substation event (GOOSE) message.

For the differential protection, the amount of uplink data is usually larger than the amount of downlink data, and the quality of service of the uplink channel has a more impact on the protection. In the uplink channel of EPON, the conflicts of packets from different ONUs are resolved by the time-division multiplexing (TDM). All ONUs are synchronized with the clock of OLT. The OLT assigns a time slot to each ONU, and the ONU sends its buffered data to the OLT when the slot arrives. The performance of EPON depends on the slot allocation scheme. Currently, the most used slot allocation scheme in EPON system for distribution networks is the interleaved polling with adaptive cycle time (IPACT) and absolute priority [16].

Once the ONU receives data from the ITU, it classifies the data frames according to the priority, pushes them into different queues, and then waits for the arrival of the time slot. When the time slot arrives with GATE message, data will be sent according to the priority, and the length of remaining data in the queues is reported to the OLT via the REPORT message. The OLT will consider the remaining queues of all ONUs and allocate the next slot length for each ONU. The IPACT with absolute priority ensures that the high-priority data can be transmitted faster, which is important for the protection. Let the maximum time slot of the ith ONU be $W_i^{\max}$, the maximum polling period is:

$$T_{\max} = \sum_{i=1}^{N} \left(G + \frac{W_i^{\max}}{R_N} \right) \tag{4.7}$$

where G is the protection interval; N is the number of ONUs; R_N is the communication speed. By changing $T_{\max}$, the delay of EPON can be controlled effectively.

When differential protection works, the data from each ITU will be sent to the server periodically. To minimize the delay in data transmission, the best way is to make $T_{\max}$ of the EPON and the transmission period of the differential protection consistent. This setting ensures that the data from ITUs can be transmitted in a polling period, and no additional waiting time needed. In this case, the time delay evaluation formula is:

$$T_{\mathrm{EPON}} = t_e + T_{\max} + t_o \times l + d/R \tag{4.8}$$

where t_e is the equipment delay caused by OLT and ONU; t_o, l, R, and d are the same definition as T_{SDH}.

4.3 Impact Analysis of EPON on Differential Protection

4.3.1 Impact Paths of EPON on Differential Protection

In EPON network, the most important factor affecting the differential protection is the time delay. Due to the limitation of EPON structure, all data should be centralized to the server. The time delay caused by optical fiber length is much larger than the delay of point-to-point structure. Besides, the polling mechanism of EPON can bring additional time delay, especially when the network is busy. Data from all ITUs must be prepared before calculating the differential algorithm, but the time delay caused by distance, polling, and the different transmission time of each ITU will make the data unable to reach the server simultaneously, causing the differential algorithm in a long waiting state.

The time synchronization error (TSE) is another important factor for CDP. The differential algorithm needs to calculate the sum of current vector from all ITUs in a protected zone. The error of time synchronization would make the sum no longer valid. If the system has a time synchronization mechanism, the time delay will put off the calculation of the sum and increase the response time of the protection. If not, the time delay will superimpose on the TSE and makes huge error to the result.

4.3.2 Impact of Time Synchronization Error

Figure 4.4 is the simplified equivalent circuit of protection zone 1 in Fig. 4.1. In this circuit, Z_0 is the positive sequence impedance (PSI) of transformer and source; Z_{L1} is the equivalent PSI of line *MM'* and load on the branch; Z_{L2} is the equivalent PSI of line and load after Q on the feeder; Z_P is the PSI of line *PM*; Z_Q is the PSI of line *NQ*; Z_M is the PSI of the line from fault point to M; Z_N is the PSI of the line

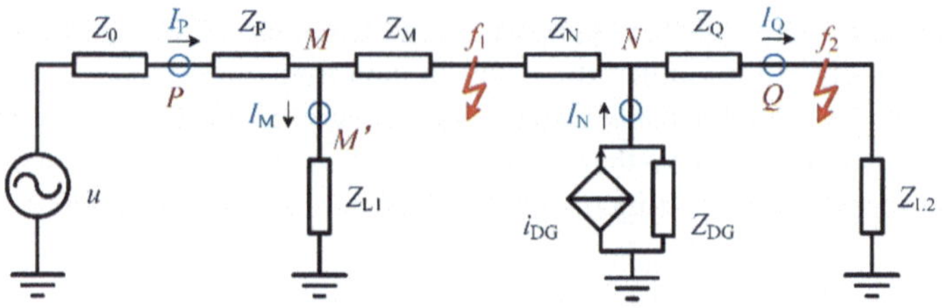

Fig. 4.4 Equivalent circuit of the distribution network with DG

from fault point to *N*; *u* is the source of the distribution network; i_{DG} is a distributed generation (DG) belonging to the IIDG; and Z_{DG} is the PSI of the DG.

Before a fault occurs, the positive sequence values of voltages and currents can be obtained after the calculation, i.e.,

$$\begin{cases} i_{P|0|} = \frac{u_{|0|}+Z_{L1}i_{|0|}}{Z_0+Z_P+Z_{L1}} \\ i_{M|0|} = \frac{u_{|0|}+Z_{L1}i_{|0|}}{Z_0+Z_P+Z_{L1}} - i_{|0|} \\ i_{N|0|} = \frac{Z_{\text{DG}}i_{\text{DG}|0|}+Z_{\text{DG}}i_{|0|}}{Z_{\text{DG}}+Z_Q+Z_{L2}} - i_{|0|} \\ i_{Q|0|} = \frac{Z_{\text{DG}}i_{\text{DG}|0|}+Z_{\text{DG}}i_{|0|}}{Z_{\text{DG}}+Z_Q+Z_{L2}} \\ u_{M|0|} = Z_{L1}i_{M|0|} \\ u_{N|0|} = (Z_Q + Z_{L2})i_{N|0|} \end{cases} \tag{4.9}$$

where $u_{|0|}$ and $i_{\text{DG}|0|}$ are the instantaneous values of *u* and i_{DG}; $i_{|0|}$ can be expressed as

$$i_{|0|} = \frac{\frac{Z_{L1}u_{|0|}}{Z_0+Z_P+Z_{L1}} + \frac{Z_{\text{DG}}^2 i_{\text{DG}|0|}}{Z_{\text{DG}}+Z_Q+Z_{L2}} - Z_{\text{DG}}i_{\text{DG}|0|}}{Z_{L1}+Z_M+Z_N+Z_{\text{DG}} - \frac{Z_{L1}^2}{Z_0+Z_P+Z_{L1}} - \frac{Z_{\text{DG}}^2}{Z_{\text{DG}}+Z_Q+Z_{L2}}} \tag{4.10}$$

When a short-circuit fault occurs at f_1, the equivalent circuit of fault component can be established, as shown in Fig. 4.5. Where $Z_{0P} = Z_0 + Z_P$, $Z_{QL2} = Z_Q + Z_{L2}$, ΔZ_{f1} is the additional impedance, $-U_{f1}$ is the virtual voltage source of fault branch. Generally, the fault current of an IIDG is determined by the control strategy and is

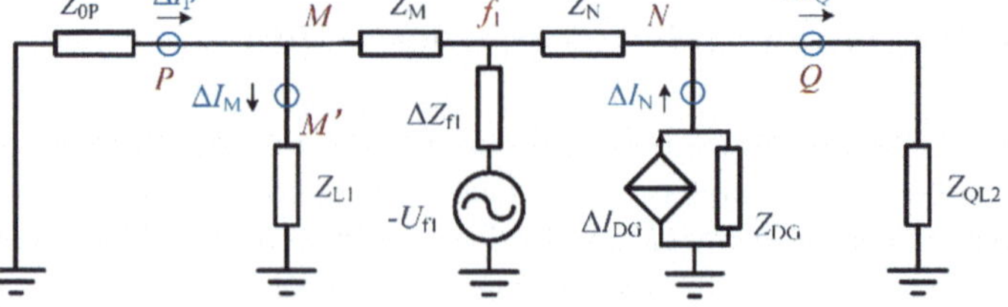

Fig. 4.5 Equivalent circuit of fault component when fault occurs at f_1

less than twice normal rating current. So in this case, ΔI_{DG} ($\Delta I_{DG} \le I_{DG}$) is used to denote the superimposed component of current supplied by DG.

The value of $U_{f1|0|}$ can be obtained by

$$-U_{f1|0|} = \left(u_{M|0|} - u_{N|0|}\right) \times \frac{Z_N}{Z_M + Z_N} + u_{N|0|} \tag{4.11}$$

The positive sequence fault component (PSFC) of fault current is

$$\Delta I_{f1} = \frac{-U_{f1|0|}}{\Delta Z_{f1} + (Z_{0M} \,||\, Z_{NL2})} \tag{4.12}$$

where $Z_{0M} = Z_M + (Z_{0P}||Z_{L1})$, $Z_{NL2} = Z_N + (Z_{DG}||Z_{QL2})$. The PSFC of each node can be derived as

$$\begin{cases} \Delta i_P = \frac{Z_{L1}Z_{NL2}}{(Z_{0P}+Z_{L1})(Z_{0M}+Z_{NL2})}\Delta I_{f1} - \frac{Z_{L1}\Delta Z_{f1}Z_{QL2}}{(Z_{0P}+Z_{L1})(Z_{0M}+\Delta Z_{f1})(Z_{0N}+Z_{QL2})}\Delta I_{DG} \\ \Delta i_M = -\frac{Z_{0P}Z_{NL2}}{(Z_{0P}+Z_{L1})(Z_{0M}+Z_{NL2})}\Delta I_{f1} + \frac{Z_{0P}\Delta Z_{f1}Z_{QL2}}{(Z_{0P}+Z_{L1})(Z_{0M}+\Delta Z_{f1})(Z_{0N}+Z_{QL2})}\Delta I_{DG} \\ \Delta i_N = \frac{Z_{0M}Z_{QL2}}{(Z_{0M}+Z_{NL2})(Z_{DG}+Z_{QL2})}\Delta I_{f1} + \Delta I_{DG} \\ \Delta i_Q = -\frac{Z_{0M}Z_{DG}}{(Z_{0M}+Z_{NL2})(Z_{DG}+Z_{QL2})}\Delta I_{f1} + \frac{Z_{0N}}{Z_{0N}+Z_{QL2}}\Delta I_{DG} \end{cases} \tag{4.13}$$

Then, the positive sequence current of node *P*, *M*, *N*, and *Q* after fault can be expressed as

$$\begin{cases} I_P = I_{P|0|} + \Delta I_P \\ I_M = I_{M|0|} + \Delta I_M \\ I_N = I_{N|0|} + \Delta I_N \\ I_Q = I_{Q|0|} + \Delta I_Q \end{cases} \tag{4.14}$$

Ignoring the DC component of the fault current, the time function of positive sequence current can be obtained by combining the current before and after the fault. The time function of each node is

$$I_{\{P,M,N,Q\}}(t)\begin{cases} I_{\{P|0|,M|0|,N|0|,Q|0|\}}(t) & (t \le 0) \\ I_{\{P|0|,M|0|,N|0|,Q|0|\}}(t) + \Delta I_{\{P,M,N,Q\}}(t) & (t > 0) \end{cases} \tag{4.15}$$

After calculating the RMS value of each node, I_{sum} and $I_{threshold}$ of the CDP can be obtained as

$$\begin{cases} I_{sum}(t) = \left|I_P(t - t_{err}) + I_M(t) + I_N(t) + I_Q(t)\right| \\ I_{threshold}(t) = S_0|I_P(t - t_{err})| + |I_M(t)| + |I_N(t)| + \left|I_Q(t)\right| \end{cases} \tag{4.16}$$

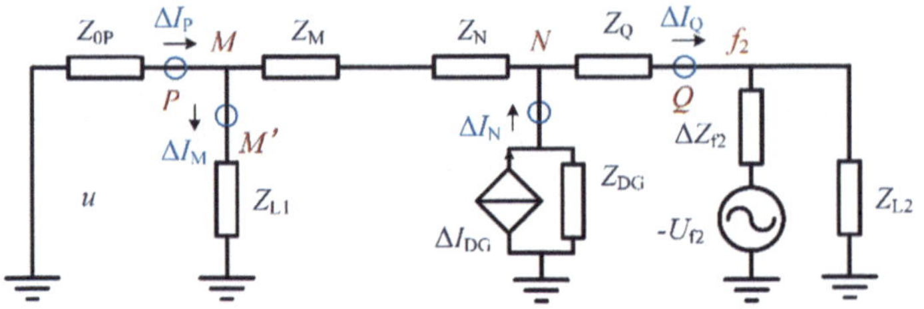

Fig. 4.6 Equivalent circuit of fault component when fault occurs at f_2

where t_{err} denotes the TSE, and the range is [0–0.02] s. Because I_P has the biggest change, it has the greatest impact on the result as described above, so only the TSE of I_P is considered in this case. Figure 4.7a shows the impacts of t_{err}.

When the fault occurs at f_2, the equivalent circuit can be obtained as Fig. 4.6 Using the same method mentioned before, the time function of positive sequence current at node P, M, N, and Q can be obtained, and I_{sum} and $I_{threshold}$ of the CDP can be calculated. Figure 4.7b shows the impacts of t_{err} in this case.

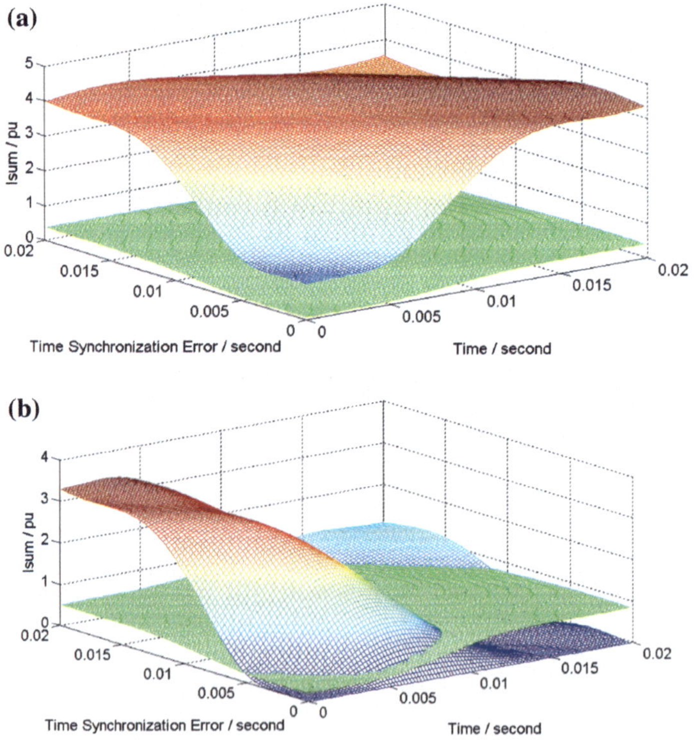

Fig. 4.7 Relationship between the error of I_{sum} and the TSE **a** fault point in the protected zone; and **b** outside the protected zone

From Fig. 4.7, one can see that the TSE has no impact on the I_{sum} when the short-circuit point is within the protected area, but has a huge impact when the short-circuit point is outside the protected zone. In Fig. 4.7b, if the data of each node is not synchronized, there will be a period that the I_{sum} is larger than $I_{\text{threshold}}$, and this period will last 10 ms in the worst case, which would trigger the protection and cut off the power supply in the non-faulty zones.

4.3.3 *Impact of Polling Period*

By using the configuration parameters in Table 4.1 [15, 16], the degree of impact of the polling period on the time delay, bandwidth, and network utilization at different network sizes can be calculated. The results are shown in Fig. 4.8.

From Fig. 4.8a, we can see that the smaller polling period can reduce the communication latency effectively. However, Fig. 4.8b, c indicates that the smaller polling period can take up a lot of bandwidth and reduce the network utilization at the same time. When 32 ITUs are connected to the EPON system, 1 ms polling period will cause 1.93 ms time delay, take up 40.96 Mbit/s bandwidth, and the network utilization is 84.9%. Comparatively, if the polling period is 0.2 ms, the corresponding results will become 1.13 ms, 204.8 Mbit/s, and 54.18%, which cannot be afforded by the existing communication network.

4.4 Modeling of Physical and Communication System

To evaluate the impact, the co-simulation combining with the physical and the communication system is an effective way. In this chapter, the co-simulation environment

Table 4.1 Parameters for communication system

System	Parameter	Setting	Unit
SDH	t_e	170	us
	t_r	100	us
	n	2	–
	l	20	km
	R	155.52	Mbit/s
EPON	t_e	150	us
	l	20	km
	R	1000	Mbit/s
Common	t_o	5	us/km
	d	160	byte

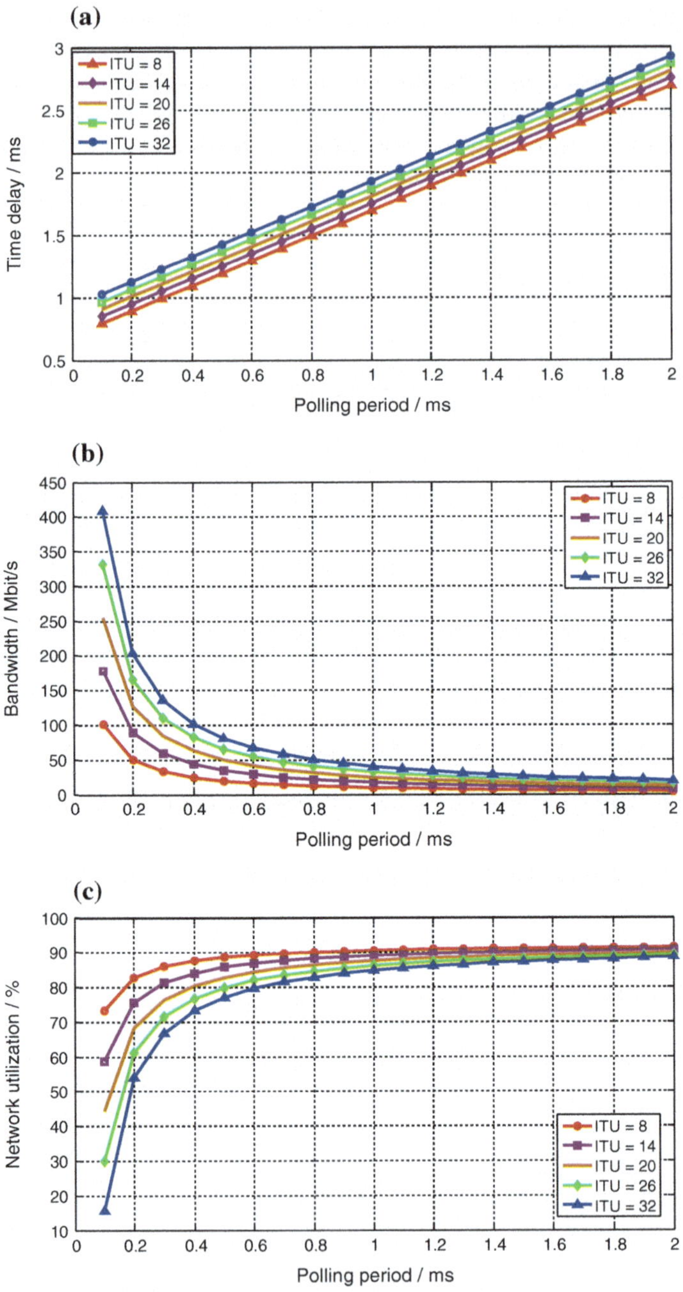

Fig. 4.8 Impact of the polling period on **a** the time delay, **b** the bandwidth, and **c** the network utilization

based on the Java agent development framework (JADE) is used for its excellent performance and scalability [17]. The model of power system is based on the modified IEEE-34 node test feeder [18], which is a long distance, light loads and uneven load distribution lines. The installation of DG in the load concentration area can effectively improve the reliability and power quality of the line [19]. In this model, two sets of distributed wind turbines DG1 (300 kW) and DG2 (300 kW) are installed at 840 and 844. When the wind turbines operate, the bi-directional current seriously affects the traditional protection system. The differential protection is a better choice for this situation. Figure 4.9 shows the network topology which is divided into 29 differential protection zones and five backup protection zones.

As the total distance of this feeder is up to 60 km, it is impossible to build a transmission channel using the EPON only. To solve this problem, a SDH transmission channel is built from substation (800) to busbar 854, and three EPON channels connect it. In this way, the length of three EPON cables can be limited in 11.1 km, 9.9 km, and 17.7 km, respectively.

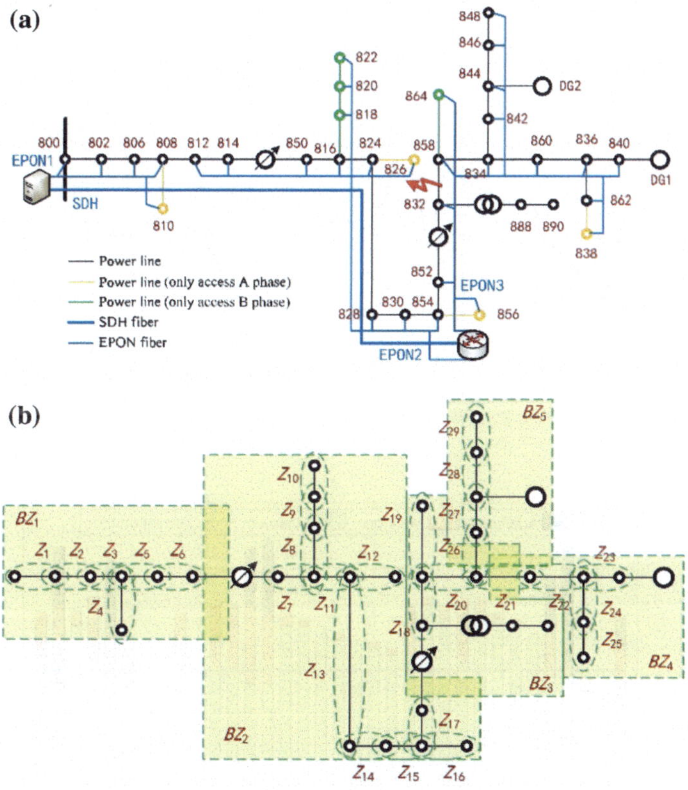

Fig. 4.9 Modified IEEE-34 node test feeder **a** network topology and configuration of the communication system; **b** protection region division

In the co-simulation model, the polling period of the communication system is set to 1 ms, and other parameters are set according to Table 4.1. S0 of the CDP is 0.1, and IS of the DCPP is 189.2 A, which is the 1.1 times of the rated current. Moreover, the tap position of the regulator affects the current value, and it is necessary to adjust the current according to the voltage ratio of the regulator.

4.5 Impact Analysis by Co-simulation

4.5.1 Case 1: Phase-to-Phase Short-Circuit Fault

A phase-to-phase short-circuit fault between A- and B-phase is triggered between busbar 832 and 858 at 0.01 s. In this case, there is no time synchronization mechanism for all ITUs, so the sampling time of each ITU is a random number within 1 ms. Figure 4.10 shows the simulation results of time delay of all ITUs and all protection zones.

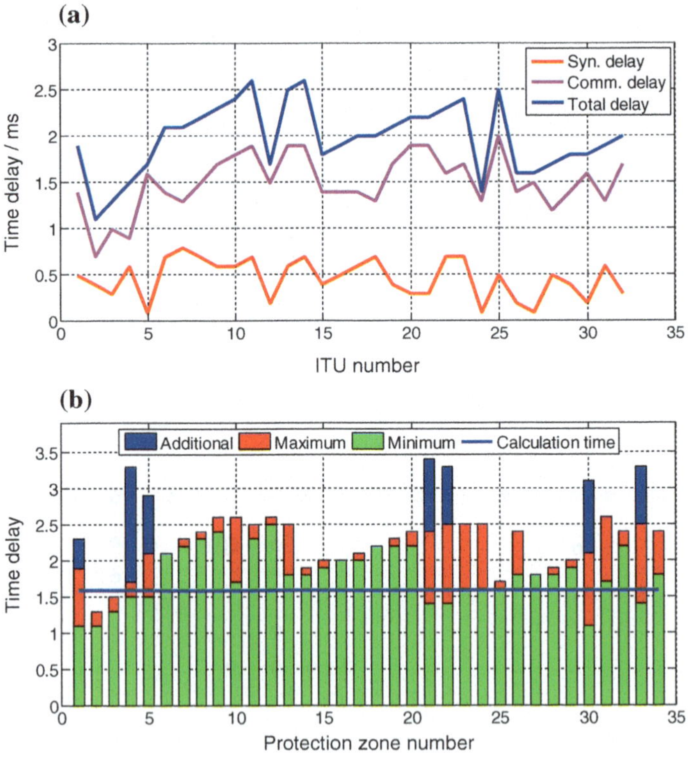

Fig. 4.10 The time delay in case 1 **a** the time delay of each ITU and **b** the time delay in each protection zone

From Fig. 4.10a, one can see that the maximum delay of these ITUs is 2.6 ms, and the minimum delay is 1.1 ms. The time lag between them will make errors. Figure 4.10b shows the impact of calculation time on the data consistency. If the maximum and minimum time delays in a protection zone are either above or below the calculation time, the data in this zone can be synchronized; but if these delays are on both sides of the calculation line, time scales of the data cannot be kept in consistency, and the additional delays will be caused which make more errors. In protection zone 1, for example, the data delays of ITU_1 and ITU_2 are 1.9 ms and 1.1 ms, respectively. When the server starts to calculate at 1.6 ms, only the data from ITU_2 is the latest, and the data one sampling period before is used for ITU_1. Because there is no time synchronization in this case, the time delay will increase from 0.8 ms to 1.2 ms.

Figure 4.11 shows the results on the responses of two kinds of differential protection schemes. In Fig. 4.11a, the proportion of fault current (PFC) is used to illustrate the judging process of the CDP, which is expressed as

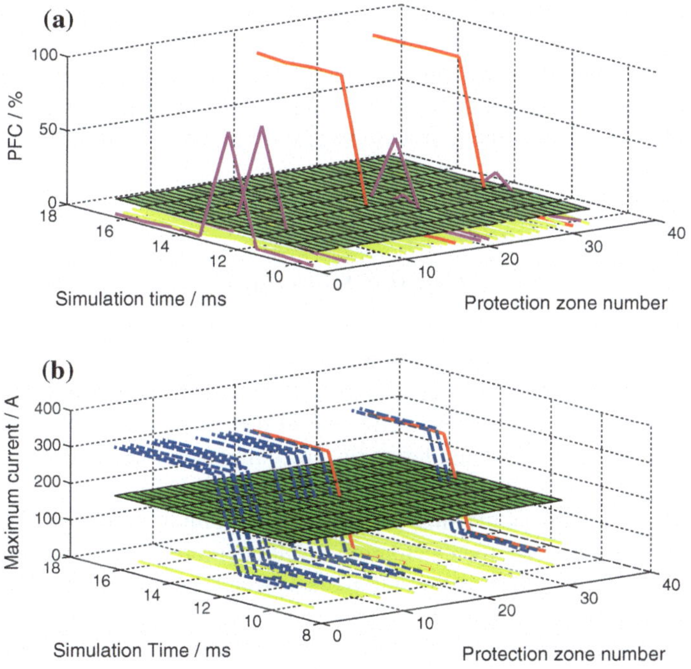

Fig. 4.11 Response of differential protection schemes for case 1 **a** PFC results and **b** maximum current of positive sequence in all zones

$$\text{PFC} = I_{\text{sum}}(t) \Big/ \sum_{k=1}^{N} |I_k(t)| \times 100\% \tag{4.17}$$

When the fault happens, the overcurrent appears in many protection zones. The maximum current is up to 322.2 A. For the CDP, the fault location (Z_{18} and BZ_3) is detected quickly, but because of the synchronization delay errors, the PFC of Z_1, Z_5, Z_{21}, Z_{22}, and BZ_4 is beyond the threshold S_0, and these zones are recognized as the fault locations as well. The reason why no errors occur in Z_4 and BZ_1 is that the ITUs with high latency in these two zones are at laterals. The current is so small that it could not affect the result. The DCPP has a better performance. In Fig. 4.11b, the overcurrent has triggered the algorithm in many zones, but only Z_{18} and BZ_3 are excluded by judging the direction of overcurrent.

From case 1, one can see that the synchronization delay has a large impact on the CDP. Since the delay can produce large value of PFC (75.1% in Z_1), it is impossible to avoid this error by raising the threshold. Moreover, only the CDP data is transmitted in this co-simulation, which means that the synchronization delay will be larger in actual applications where other data, from the supervisory control and data acquisition (SCADA) system, or the voice over internet protocol (VOIP) system, will be added to transmit. Comparing to the CDP, the DCPP is not sensitive to the information system error and can recognize the fault location correctly. Therefore, the DCPP is a better choice for locating the fault with large current when the information system is not good enough.

4.5.2 Case 2: Phase-to-Ground High-Impedance Fault

A phase-to-ground high-impedance fault is triggered between busbar 832 and 858 at 0.01 s. The fault occurred at phase-A, and the fault resistance is set as 100 Ω. The time synchronization mechanism is effective in this case, so there is no sampling time delay for each ITU. The simulation results of time delay are shown in Fig. 4.12.

From Fig. 4.12, one can see that the most delay of all ITUs is decreased, and the delay of all zones displays the similar change. The point to emphasize here is that these delays will not affect the calculation result anymore because the sampling is global synchronization, and the timestamp of each data is used to protect the old data from being overwritten by new one.

Figure 4.13 shows the simulation results on the response of two differential protection schemes. By using the time synchronization technology, the CDP locates the fault correctly, and the small fault current does not affect the recognition of the CDP. In this case, the DCPP cannot protect the network because all currents are below the threshold I_s.

This case demonstrates the sensitivity of the CDP effectively. From Fig. 4.13a, one can see that the CDP is effective even if the fault current decreases further. In contrast to the CDP, the DCPP is unable to recognize the high-impedance fault

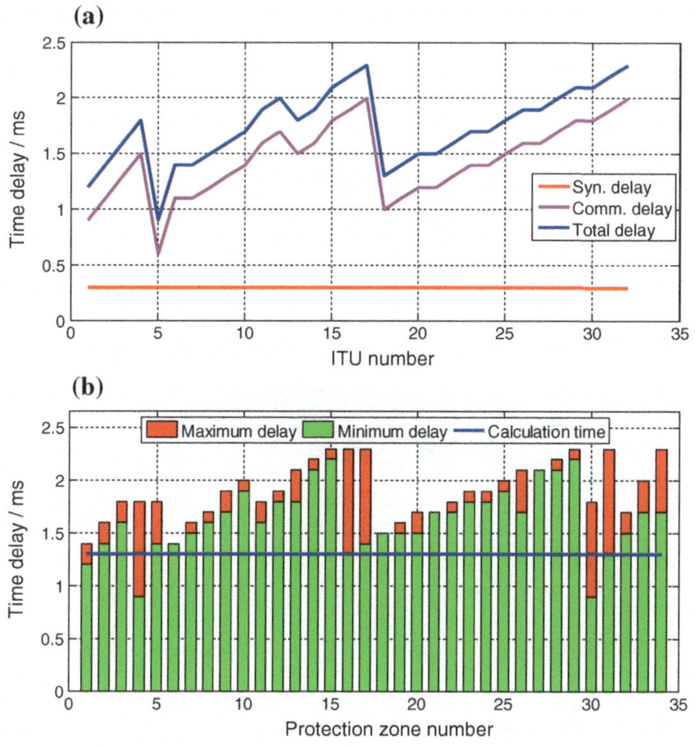

Fig. 4.12 Time delay in case 2 **a** the time delay of each ITU and **b** the time delay in each protection zone

current. Although threshold reduction can improve the sensitivity of the DCPP, the probability of erroneous judgment will also rise because the DCPP is impossible to distinguish the reason of the increased current caused by fault or the load increases in normal operation.

4.6 Summary

This chapter summarizes the principles of two typical protection algorithms for distribution networks, designs the calculation process of differential protection based on EPON, analyzes the impacts of EPON-based communication networks on the differential protection, and verifies the feasibility and performance by the co-simulation of a modified IEEE-34 test feeder. Theoretical analysis and co-simulation results show that the differential protection based on EPON can protect the distribution network effectively. In the system with time synchronization, CDP can achieve precise fault location and isolation, while DCPP has better reliability in the system without time

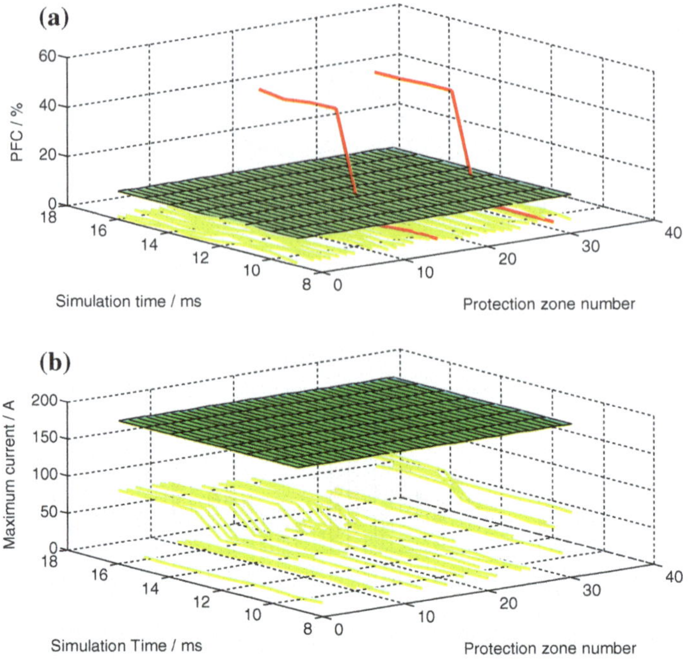

Fig. 4.13 Response of differential protection schemes for case 2 **a** PFC results and **b** maximum current of positive sequence in all zones

synchronization. The co-simulation also provides a guideline for smart distribution networks to select a suitable differential protection scheme.

References

1. Thirumalai A, Liu X, Karady GG (2011) Ultra-fast pilot protection of a looped distribution system. In: PIEEE PowerTech, Trondheim, Norway, 19–23 June 2011
2. Gao H, Li J, Xu B (2017) Principle and implementation of current differential protection in distribution networks with high penetration of DGs. IEEE Trans Power Del 32(1):565–574
3. Han B, Wang G (2016) An improved pilot protection for distribution network with inverter-interfaced distributed generations. In: IEEE PES Asia-Pacific power and energy conference, October 2016
4. Tang J, Gong Y, Schulz N, Steurer M (2008) Implementation of a ship-wide area differential protection scheme. IEEE Trans Ind Appl 4(6):1864–1871
5. Xu S, Liu Y, C Cai (2012) Study of novel sine fitting differential protection algorithm in distribution system with high penetration of DGs. In: IEEE Power and Energy Society General Meeting, July 2012
6. Xu M, Zou G, Xu C, Sun W, Mu S (2016) Positive sequence differential impedance protection for distribution network with IBDGs. In: IEEE international conference on power system technology, September 2016

7. Nafi NS, Ahmed K, Datta M, Gregory M A (2014) A novel zigbee based pilot protection scheme for smart distribution grid. In: Australasian telecommunication networks and applications conference, November 2014
8. Eissa MM (2012) Protection technique for complex distribution smart grid using wireless token ring protocol. IEEE Trans Smart Grid 3(3):1106–1117
9. Yu P, Xi H, Qiang Y (2012) The research of interoperability of EPON systems in power systems. In: International conference on advanced power system automation and protection, April 2012
10. Sun Z, Ma Y, Sun F, Wang Y (2010) Access control for distribution automation using ethernet passive optical network. In: Asia-Pacific power and energy engineering conference, March 2010
11. Ahmed MA, Kim Y (2013) Protection schemes of EPON-based communication network for offshore wind power farms. In: International conference on ICT convergence, October 2013
12. Lévesque M, Maier M (2012) The Über-FiWi network: QoS guarantees for triple-play and future smart grid applications. In: 14th international conference on transparent optical networks, July 2012
13. Tang F, Zha X (2012) Reliability analysis of smart distribution grid communication system based on EPON. In: Asia-Pacific power and energy engineering conference, March 2012
14. Kim K, Chung H (2016) Real-time demonstration of extended 10G-EPON capable of 128-way split on a 100 km distance using OEO-based PON extender. In: International conference on information and communication technology convergence, October 2016
15. Zhao D, Sun Y (2002) A preliminarily analysis of characteristics of code error and time delay of protective relaying signals transmitted by SDH optical fiber communication system. Power Syst Technol 26(10):66–70
16. Kramer G (2005) Ethernet passive optical networks. McGraw-Hill Education - Europe, New York
17. Duan Y, Luo L, Li Y, Cao Y, Rehtanz C, Markus Küch (2017) Co-simulation of distributed control system based on JADE for smart distribution networks with distributed generations. IET Gener Transm Distrib 11(12):3097–3105
18. IEEE 34 Node Test Feeder. IEEE PES AMPS DSAS test feeder working group. [Online]. Available: http://sites.ieee.org/pes-testfeeders/
19. Adewole AC, Tzoneva R (2012) Fault detection and classification in a distribution network integrated with distributed generators. In: IEEE PES PowerAfrica 2012 conference and exhibition, July 2012

Chapter 5
Modeling and Simulation of Data Flow for VLAN-Based Substation Communication System

The digitalization and networking of secondary equipment for substation have been applied in power system. Based on IEC 61850, the standard for communication in substations has resolved the interoperability between intelligent electronic devices (IEDs) from different vendors. Thus, the IEDs' uniform communication standard contributes to the realization of more sophisticated functionality in substation automation systems, which put forward the higher requirement on real-time capability and reliability of the whole automation system. However, there are still several issues unsolved for the simulation, planning, and extension of substation communication network (SCN).

In this chapter, three kinds of mathematical models are proposed for typical data flow within substations according to IEC 61850, which are cyclic data, stochastic data, and burst data. Thereby, a quantitative analysis of data flow is carried out for a typical substation based on the proposed data models. The advantage of VLAN and impacts of system faults as well as network topologies on a VLAN-based network are also evaluated and simulated by OPNET Modeler. The data flow models are beneficial for the acquisition of more convincing results to assess network performance. Thus, the simulation results for a sample substation can be used to support power utility personnel with the planning and construction process of SCN.

5.1 Introduction of VLAN Technology

Substation automation systems (SASs) have entered a new stage of digitalization and intelligence with the development of advanced communication technologies and intelligent electronic devices (IEDs). The IEC 61850-based substation has broken the technical bottleneck of traditional SAS, which facilitates the effective substation monitoring, local and remote control, protection, and primary equipment condition monitoring and more importantly resolves the interoperability of different IEDs [1].

According to IEC 61850, one of the main characteristics for smart substation is the digitalization of secondary equipment, which is implemented by changing

Y. Cao et al., *Cyber-Physical Energy and Power Systems*,
https://doi.org/10.1007/978-981-15-0062-6_5

the traditional electromagnetic current transformer and potential transformer into optical or electronic CT and PT, as well as replacing the copper wire with optic fiber. Consequently, communication network has permeated SAS more deeply than ever before, so the real time, reliability, and security of substation communication networks (SCNs) must be ensured under any operating conditions and contingencies in order to maintain the normal operation of substation.

Although there has been much work on studying the substation communication system by means of LAN simulation, there are still several issues unresolved in order to assess the dynamic performance of SCN comprehensively and accurately. The enhancement of the previous research is required for two reasons: (1) There are no proper mathematical models to describe various messages in SAS, which are the basics for accurate evaluation of network performance by using simulation tools, such as OPNET Modeler; (2) the comprehensive assessment of real-time performance based on the proposed data flow models is needed in order to acquire more convincing results. In fact, specific mathematical models for data flow were not established, and the traffic management technologies, such as VLAN, were not thoroughly evaluated [2].

VLAN is a network management technology based on LAN switching to realize the flexible control of data broadcast range, which logically forms isolated subnets in the same local area network (LAN) by dividing network devices into different partitions. The VLAN partitioning is not limited by the physical location of devices. Users can configure the switch's ports or routers according to their own needs to achieve logical grouping of multiple devices. Thus, without changing the physical structure of the network, a LAN can be divided into multiple logical subnetworks.

When broadcasting data in LAN, the devices located in the same logical subnet can receive data from each other, while devices in different logical subnets cannot communicate with each other directly. Therefore, VLAN can limit the broadcasting range of different data in LAN and improve the security of key data transmission and the availability of network bandwidth.

The principle of VLAN technology is illustrated in Fig. 5.1, which compares the data flow changes in a simple network before and after VLAN configuration. In Fig. 5.1a, the network has not been configured with VLAN policy, the data packets transmitted from device 1 to device 4 by broadcasting mode. Since devices 1–5 are connected to the same LAN, other devices in the LAN can also receive the broadcasted data from device 1. However, in Fig. 5.1b, the five devices in the LAN are divided into three logical subnets by configuring the VLAN. In this scenario, device 1 and device 4 are in the same subnet, whose VLAN ID is 1, while device 2 and device 5 are in another subnet, with VLAN ID 2, and the remaining device 3 constitutes a single subnet with VLAN ID 3. When device 1 transmits data to device 4, each data packet has a unique VLAN ID, and the data packets that are inconsistent with the VLAN ID preset will be automatically filtered out when they are forwarded through the switch. The broadcasted data packets can be forwarded by switch port connected with device 4, and there is no data communication between other devices and switches.

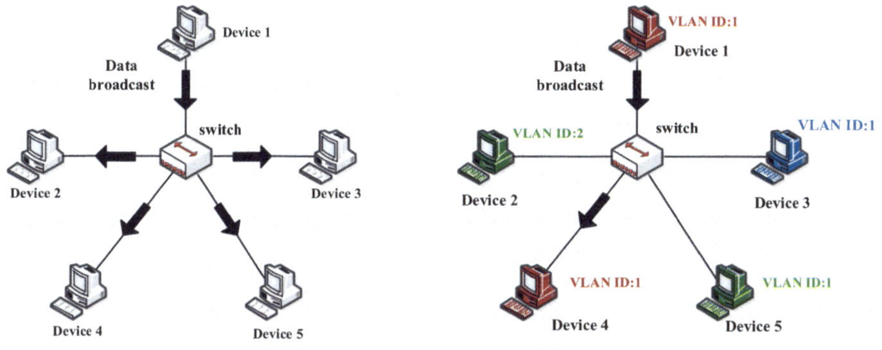

(a) Data flow before VLAN configuration (b) Data flow after VLAN configuration

Fig. 5.1 Data flow change before and after VLAN configuration

It is obvious that, without changing physical structure, the amount of redundant data transmitted in the network can be reduced via reasonable VLAN configuration, which improves the utilization of network bandwidth. At the same time, the security of data transmission in the network can also be improved through logical isolation of LAN devices.

5.2 Theoretical Models of Data Flow

According to IEC 61850-5 [3], all messages in a substation are divided into seven types, which are *fast message*, *medium-speed message*, *low-speed message*, *raw data message*, *file transfer function*, *time synchronization message*, and *access control command*.

However, such classification has some kind of redundancy. For example, the sampled values generated by IEDs belong to *raw data message*, which are also *fast message* type. Therefore, from the perspective of data flow characteristics in time domain, the messages in substation have been divided into three types in this chapter that is cyclic data, stochastic data, and burst data.

5.2.1 Modeling for Cyclic Data Flow

The cyclic data flow is generated by periodic sampling of CT and PT. According to the practical operation condition of substations, there are two types of cyclic data. One type is the sampled values (SAVs) generated by merging unit (MU) IEDs in substation process level and then transmitted to protection and control (P&C) IEDs in substation bay level [4]. SAVs represent time-critical information that contains large amounts of data flow and will have an intensive influence on the SAS network.

The other type is the meter values and breaker status information transmitted from the device in bay level to the server in station level at a certain time interval, which belong to cyclic type of Generic Object-Oriented Substation Event (GOOSE) message [5]. This kind of cyclic GOOSE data is comparatively stable and a *medium-speed message* type.

Cyclic data is a typical time-driven data of fixed length, which means that messages are triggered at the same time intervals and the packet size can be decided in advance. Thus, cyclic data can be modeled as follows [6].

$$M_c = f(L_c, N_c, D_c) \tag{5.1}$$

$$N_c = f_0 \tag{5.2}$$

$$D_c = S_c + E_c + R_c \tag{5.3}$$

where L_c is the size of cyclic data, which contains frame header, address field, data field, CRC field, and so on; N_c is the number of cyclic data arrived per unit time, numerically equal to the sampling frequency f_0 of IEDs; D_c is the time delay of a message from end to end, representing the sum of Ethernet delay E_c, pre-treatment time of the sender S_c, and post-proceeding time of the receiver R_c.

According to IEC 61850-5 [3], SAVs should be transmitted and processed within $D_s = 3$ ms for class P2 and P3; otherwise, the system's performance will be influenced which may have disastrous consequences. For example, if there is a short-circuit fault in the substation, the protection IEDs will not act immediately without the timely transmission of SAVs, which may cause electrical equipment to be out of service and loss of power loads. Therefore, the end-to-end delay for SAVs should satisfy the following constraint.

$$D_c \leq D_s \tag{5.4}$$

Similarly, the end-to-end time delay for breaker and equipment status information, which belongs to *medium-speed message* type, should satisfy Eq. (5.4) with $D_s =$ 100 ms.

The flow packet generation diagram of cyclic data is shown in Fig. 5.2.

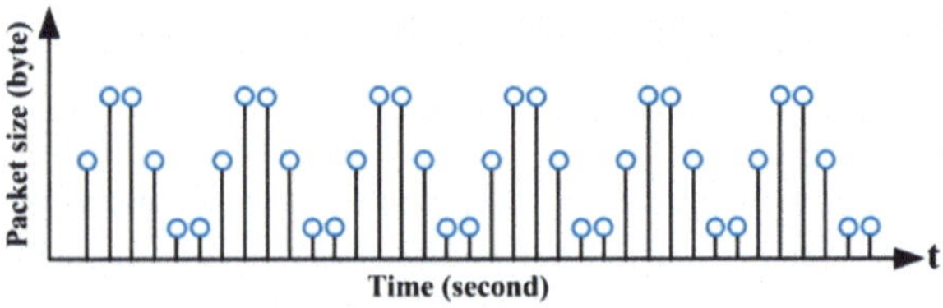

Fig. 5.2 Generation of data packets for cyclic data flow

5.2.2 Modeling for Stochastic Data Flow

Stochastic data is a typical event-driven data, which means it is triggered by accidents or unplanned events, such as the trip message when a short-circuit fault occurs and the artificial modulation of equipment parameters. Stochastic data in substations can be mainly divided into two types:

(1) *Type* 1: Transformer tap modulation, switch operation message, trip message, protection function interlocking, time synchronization, capacitor switching, and so on.

It usually has the features of small size and short duration, while the transmission time should meet the requirements of *fast message* type.

(2) *Type* 2: Protection setting modification, event log checking, recording data transmission, file transfer, and so on.

Type 2 is large in size and will usually cause a sudden increase of network flow, while the real-time requirement of transmission is not strictly specified.

Generally, stochastic data has the following characteristics of time sequence: The packet is generated in a random time period with the probability of P. The size of the packet can be fixed or time-variant. There is no correlation between two packets arriving one after the other, which means that the amount of packets in two mutually exclusive time periods is independent. Therefore, the arrival of stochastic data can be modeled by the Poisson process.

For the time period $[\tau, \tau + t]$, supposing λ is the average arrival rate of packets (the number of packets arrived per unit time), $N(t)$ represents the total number of arrived packets. The probability of k packets arrived in time interval t will obey the Poisson distribution with parameter λ:

$$P\{N(\tau + t) - N(\tau) = k\} = \frac{(\lambda t)^k e^{-\lambda t}}{k!} \tag{5.5}$$

While the probability of received packets obeys the Poisson distribution with parameter of λ, the time interval between two consecutive packets will obey negative exponential distribution with parameter of $1/\lambda$, which means the average time interval of packets arrived is $1/\lambda$. The probability density function for negative exponential distribution is defined by:

$$g(t) = \lambda e^{-\lambda t}, t \geq 0 \tag{5.6}$$

According to IEC 61850-8-1, trip message is required to send by the heartbeat mode with a changing time interval. So, the parameter $1/\lambda$ is necessary to simulate the generation of such stochastic data types. A more detailed description can be found in Sect. 5.3.

The stochastic data flow packet generation diagram is shown in Fig. 5.3.

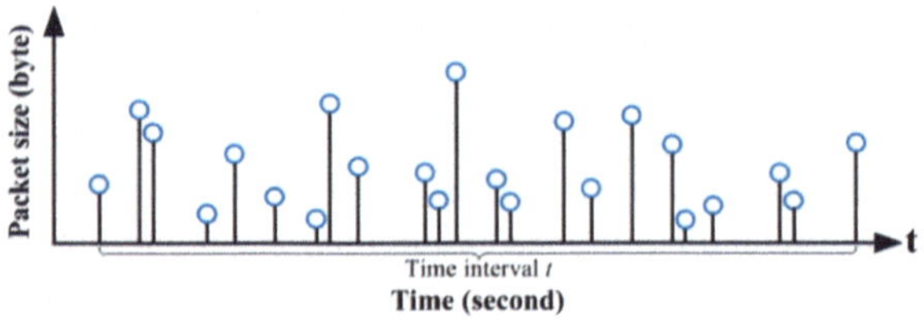

Fig. 5.3 Generation of data packets for stochastic data flow

5.2.3 Modeling for Burst Data Flow

During a random time, burst data is not only generated with the probability of λ, but also dependent on the previously occurred events. Burst data mainly contains information about protection actions and the changing status of breakers, which belong to GOOSE message too. When a fault occurs, the protection device acts and then the transmission of GOOSE message is changed from cyclic mode to burst mode [5], which consequently generates burst data flow.

In burst mode, the transmission interval for GOOSE message changes from T_0 to $T_i = 2^i \times S(i = 1, 2, 3, \ldots)$, where T_0 is the constant time interval for GOOSE message in normal state and T_i represents the transmission interval of ith GOOSE packet since trip message generates. The reduplicative transmission for this piece of GOOSE message ends until T_i reaches T_0. The values for T_0 and S are set in substation configuration description (SCD) file [5]. To avoid confusion, the GOOSE message mentioned below refers to burst-type GOOSE message particularly, while the cyclic GOOSE message corresponds to meter values and breaker status information in normal condition.

Generally, burst data will cause a large data amount on the network in a relatively concentrated transmission time. The arrival of burst data packets has characteristics of time after effect, which means there appears a short time period of data transmission on SAS network when the burst data is generated, while the network is free for a long period of time after the transmission of data packets. Therefore, this type of data flow has the characteristics of long-range dependence and self-similarity, which presents the same burstiness at different timescales [7, 8].

The burstiness, long-range dependence, and self-similarity of Ethernet data flow have been generally accepted by researchers [9–11]. It has been proven that heavy-tailed distribution and *ON/OFF* model can be used to describe the self-similarity of network data flow [12–14]. In an *ON/OFF* model, it is assumed that the data source states change repeatedly between sending and not sending messages. When the state is *ON*, data is generated with a constant rate, whereas none is generated when the state is *OFF* [9]. Generally, consequent *ON* and *OFF* states are independent and identically distributed. Therefore, it is applicable to describe the characteristics of the *ON/OFF* model by setting the distribution of time duration for both states.

Supposing time duration of *ON* state for a single data source obeys the Pareto distribution, which is a typical heavy-tailed distribution. The cumulative distribution function of Pareto distribution can be described by

$$F(t) = P(T \le t) = 1 - \left(\frac{k}{t}\right)^{\alpha}, \quad 0 < k \le t, \alpha > 0 \tag{5.7}$$

where k is the minimum possible value of T, which represents the minimum duration of *ON* state; α is a positive parameter. The Pareto distribution is characterized by a scale parameter k and a shape parameter α known as the *tail index* [15].

The mean value for Pareto distribution is shown as follows:

$$E(T) = \begin{cases} \infty & \text{if } \alpha \le 1, \\ \frac{\alpha k}{\alpha - 1} & \text{if } \alpha > 1. \end{cases} \tag{5.8}$$

The time duration for OFF state obeys negative exponential distribution of Poisson process, with the same probability density function shown in (5.6). As long as the duration of ON or OFF state obeys heavy-tailed distribution, the accumulation of large numbers of these ON and OFF states will form self-similar burst data flow [9, 14]. There is only one parameter H, so-called Hurst coefficient, in the self-similar data model. Essentially, the Hurst coefficient describes the degree of self-similarity of data flow as well as the decaying rate of the self-similar function. H satisfies the following equation [9]:

$$H = \frac{3 - \alpha_{\min}}{2} \tag{5.9}$$

where $\alpha_{\min}$ is the smaller one of parameter α for heavy-tailed distribution between ON and OFF states. For a typical self-similar sequence that represents the network data flow, the range of H is 0.5–1. The degree of self-similarity will grow when the Hurst coefficient H increases from 0.5 to 1.

Obviously, parameter α should satisfy $1 < \alpha < 2$ with H limited to 0.5–1. Therefore, the Pareto distribution will have finite mean value and infinite variance according to Eqs. (5.8) and (5.9). The burst data flow packet generation diagram is shown in Fig. 5.4.

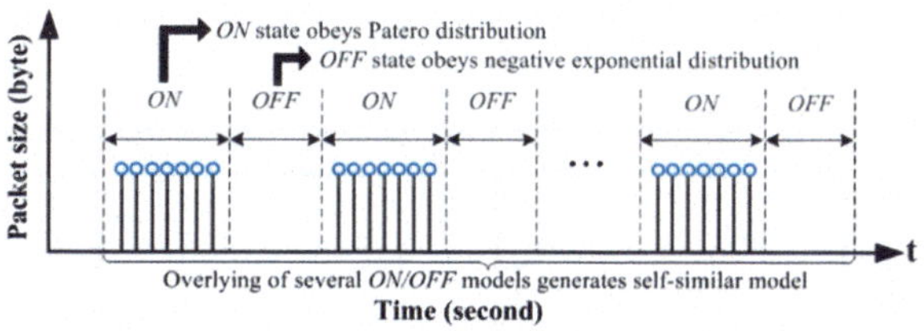

Fig. 5.4 Generation of data packets for burst data flow

5.3 Analysis of Data Flow in a Substation

5.3.1 Typical Structure for Substation System

Based on the theoretical models proposed in Sect. 5.2, a detailed analysis of data flow for a typical substation has been carried out in this section. The amount of data for five types of messages is calculated, respectively. There are three different types of IEDs in a typical SAS [16], which are MU IED, breaker IED, and P&C IED. Firstly, the MU IED processes and combines the signals from field current transformer (CT) and voltage transformer (VT). Then, it transmits the digital voltage and current output to the process bus. The breaker IED not only controls the breaker's position but also monitors its state and condition. Thereby, it receives the switching commands from the P&C IEDs and sends state change event messages to corresponding protection IEDs through the process bus. The P&C IED, a multi-functional device, bundles the protection and control functions for the bay unit it serves within the substation.

All the messages in a substation are transmitted between either the IEDs described above or the IEDs and the station server. Thereby, a typical 220 kV D2-1-type substation with two transformer bays (T1 and T2), one bus section bay (S), and six feeder bays (F1–F6) has been studied [3]. Figure 5.5 shows the single-line diagram and physical bays of the substation. Note that every transformer bay contains two MU IEDs, one P&C IED, and two breaker IEDs; the bus section bay contains one MU

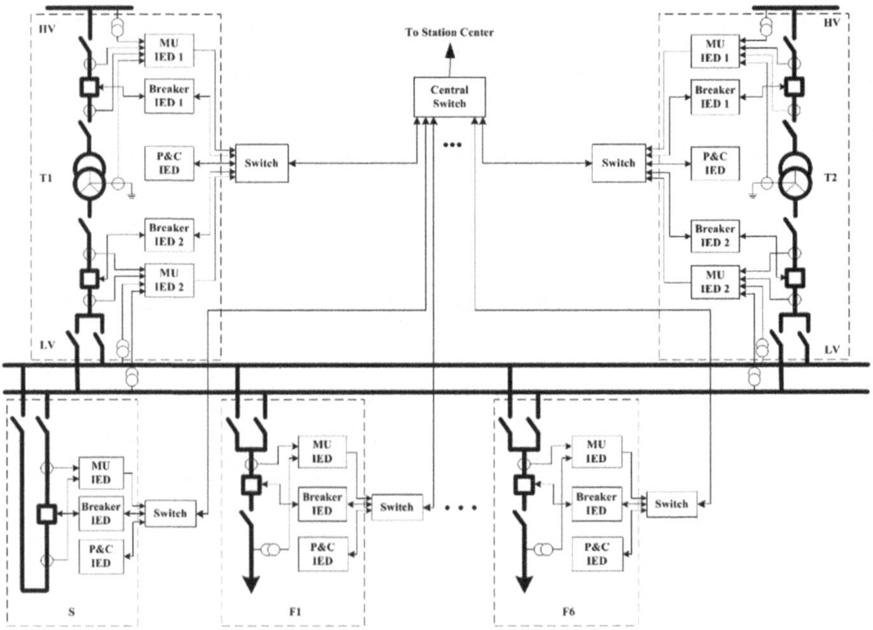

Fig. 5.5 Single-line diagram for a typical 220 kV D2-1-type substation

IED, one P&C IED, and one breaker IED; the feeder bay contains one MU IED, one P&C IED, and one breaker IED.

5.3.2 Data Flow for Substation Communication System

Analysis of data flow, which is the premise and foundation of SAS network simulation, has been carried out for a VLAN-based substation displayed in Fig. 5.6. Its network structure and communication traffic correspond to the single-line diagram of Fig. 5.6. The SAVs will be transmitted by means of broadcasting if there is no filtering scheme, which means P&C IEDs in a certain bay will receive SAVs from MU IEDs in every bay of process level if there is no VLAN and multi-cast configured. Consequently, the broadcast of SAVs in a SCN without VLAN will cause large amounts of redundant messages, which may affect the communication efficiency of substation and cause the malfunction of devices. However, in fact, the P&C IEDs in a certain bay only need the SAVs transmitted from MU IEDs inside the same bay. Thus, the VLAN scheme is adopted and studied to limit SAV messages in a particular bay.

(1) *Messages from MU IED to P&C IED*

MU IEDs in every bay send SAVs to P&C IEDs within the same bay at a certain rate. The data is transmitted by means of broadcast without filtering. According to Sect. 5.2, SAV is a typical cyclic data flow with fixed message size. Figure 5.7 presents the ISO/IEC 8802-3 frame format for SAVs as well as the ASN.1 coded APDU frame structure according to IEC 61850-9-2LE.

It is supposed that the Application Protocol Data Unit (APDU) of SAVs sent by MU IED in each bay has two Application Service Data Units (ASDUs). According to IEC 61850-9-2LE, the data set of ASDU comprises four voltages and four currents (three phases and neutral for each). Thus, the typical size for a single ASDU is 93 bytes with 64 bytes of measurements. As shown in Fig. 5.7, to add the 802.1 Q header and savPdu, noASDU, sequence of ASDU, etc., the packet size for SAVs with two

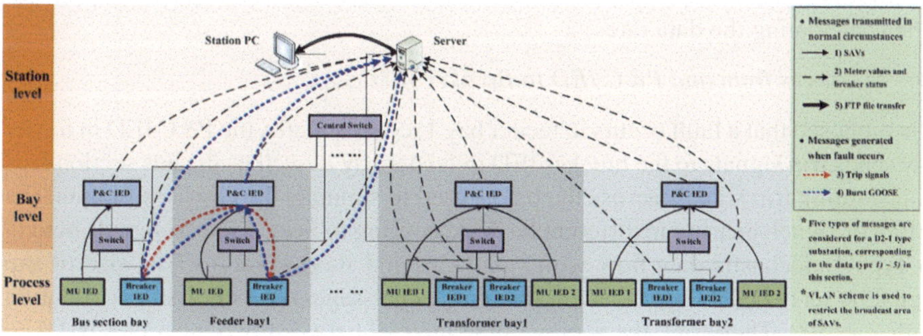

Fig. 5.6 Data flow diagram for a VLAN-based D2-1-type substation

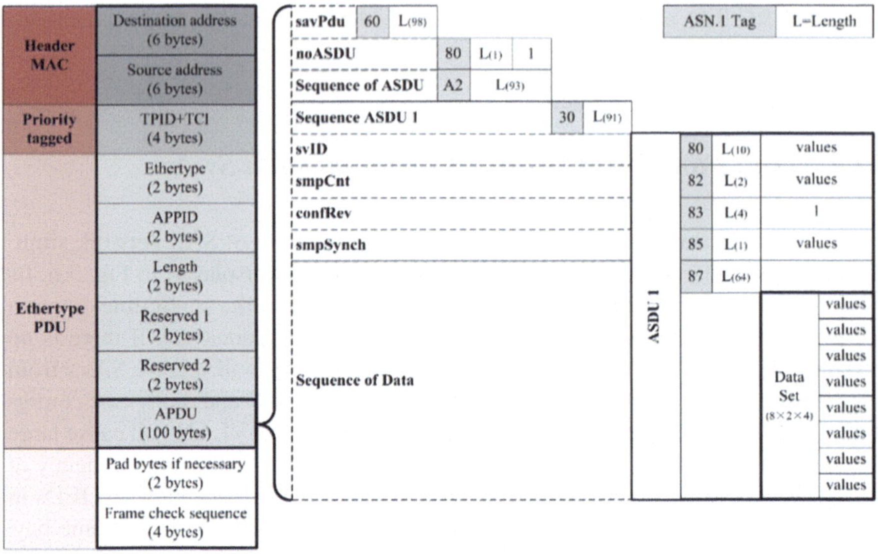

Fig. 5.7 Frame format for SAV packet

ASDUs is 219 bytes in total. Meanwhile, the preferred sampling rate for SAVs is 4800 Hz with two ASDUs. Thus, it is easy to figure out that the data rate for MUs is 2400 packets per second with two ASDUs per frame, and then the data rate from MU IED to P&C IED in each bay amounts to $T = 2400$ packet/s $\times$ 219 byte/packet $\times$ 8 bit/byte = 4.2048 Mbit/s.

(2) *Messages from Breaker IED and P&C IED to Server*

In normal condition, the breaker IEDs and P&C IEDs in all bays send meter values and breaker status information (cyclic GOOSE) to the server with a constant time interval of 20 ms and the message size is set to 144 bytes [17]. Typically, these messages also belong to cyclic data flow type as well as medium-speed messages that are mapped to MMS protocol suits which have a TCP/IP stack above the Ethernet layer [16]. Thus, it is required to add the size of header and tail of Ethernet protocol when calculating the data rate.

(3) *Messages from and P&C IED to Breaker IED*

It is supposed that a fault occurs in feeder bay 1; consequently, the P&C IED in feeder bay sends trip signals to the breaker IED in feeder bay as well as the bus section bay. Transmitting trip signals across the bay is used to simulate the case of simultaneous action of switch gearing in different bays. According to Sect. 5.2, trip signals belong to Type 1 stochastic data flow with small size and short duration. The size of trip signals is set to be 204 bytes, and the arriving of messages obeys Poisson distribution with $\lambda = 500$. Therefore, the average time interval between two messages is $1/\lambda = 2$ ms. According to IEC 61850-8-1, after the first transmission, this type of message needs to be sent repeatedly in a time sequence of $2n(n = 1, 2, 3, \ldots)$ ms. So, it is the extreme case considered in simulation study with the setting of $1/\lambda = 2$ ms.

(4) *Messages Generated after Transmission of Trip Signals*

Once the breaker IED acts based on the received trip message, it will send a GOOSE message to the corresponding P&C IED as well as the server. Meanwhile, the P&C IED will send a GOOSE to the server too. According to Sect. 5.2, these messages are typical type of burst data flow. For the ON state of burst data, the data size is set to be 144 bytes and the arriving time interval is 0.1 ms. The performance of ON state obeys the Pareto distribution with typical parameters of $k = 512$ μs and $\alpha = 1.1$, while the generation of OFF state obeys the Poisson distribution with $\lambda = 263.16$. According to (5.8), it can be concluded that the average duration time for ON state is 6.2 ms, while that of OFF state is 3.8 ms.

(5) *Messages from Server to Station PC*

The FTP server is used to send a large amount of data to station PC in a randomly selected time, which indicates that messages from server to station PC belong to Type 1 stochastic data flow. The total data amount for file transfer is supposed to be hundreds of megabytes. Assume that the single packet size for FTP application is 1000 bytes and the time interval for consecutive packets obeys exponential distribution with parameter of $1/\lambda = 1$ ms. Thus, the average data rate for file transfer is 8 Mbit/s.

A summary of five typical types of messages transmitted in D2-1 substation is shown in Table 5.1.

(1) The transmission of SAVs, together with meter values and breaker status information, is started at 10 s. Both of them are modeled as cyclic data flow and last until the end of simulation.
(2) There happens to be a fault at 20 s in feeder bay F1, and the P&C IED sends trip signals to breaker IED in F1 and bus section bay S once the fault occurs.
(3) Trip signals are modeled as stochastic data flow and transmitted repeatedly between 20 and 20.3 s. The protection IEDs act once trip signals are sent; thus, the transmission of GOOSE is set to begin at a random time between 20 and 20.01 s. GOOSE messages are modeled as burst data flow in OPNET and last for 300 ms.
(4) There has been a file transfer in the SCN when the fault occurs. The FTP file transfer is modeled as stochastic data flow, which begins at 18 s and lasts until the end of simulation because of the large data amount.

The SAVs are time-critical messages and therefore can be directly mapped to data link layer of the TCP/IP model to reduce the expense of high-layer protocol [17]. The three-layer *ethernet_station* node in OPNET is used to model MU IEDs, and generation of cyclic SAVs is implemented by properly setting the parameters of the node. Trip signals and GOOSE messages belong to *fast message* type and thus are directly mapped to data link layer too. The *ethernet_workstation* node is used to model P&C and breaker IEDs in simulation.

Table 5.1 Summary of data flow transmitted in typical substation

Message name	Source and destination ($i = 1, 2, \ldots, 6; j = 1, 2$)		Data type	Packet size (bytes)
	Without VLAN scheme	With VLAN scheme		
SAVs	S_MU IED → S_P&C IED S_MU IED → F_P&C IEDs S_MU IED → T_P&C IEDs Fi_MU IED → S_P&C IED Fi_MU IED → F_P&C IEDs Fi_MU IED → T_P&C IEDs Tj_MU IED → S_P&C IED Tj_MU IED → F_P&C IEDs Tj_MU IED → T_P&C IEDs	S_MU IED → S_P&C IED Fi_MU IED → Fi_P&C IEDs Tj_MU IED → Tj_P&C IEDs	Cyclic data	219
Meter values and breaker status	P&C_IEDs → server Breaker_IEDs → server		Cyclic data	144
Trip signals	F1_P&C IED → F1_Breaker IED F1_P&C IED → S_Breaker IED		Stochastic data	204
GOOSE	F1_Breaker IED → F1_P&C IED S_Breaker IED → F1_P&C IED F1_Breaker IED → server S_Breaker IED → server F1_P&C IED → server		Burst data	144
FTP file transfer	Server → station PC		Stochastic data	1000

In addition, the application of file transfer, which has a large amount of data but relatively low transmission time limit, is implemented by the *server* node. The TCP/IP protocol is used for reliable transmission. A summary of communication modes, protocols, and adopted models for different types of messages is shown in Table 5.2.

Based on the data flow models proposed in Sect. 5.2, four types of applications and profiles are configured in simulation, corresponding to trip signals, breaker status information, GOOSE messages, and FTP file transfer. The transmission of the first three messages is all modeled by *videoconference application* with a different *frame size* and *inter-arrival time*. FTP file transfer is modeled by *FTP application*. Profiles corresponding to these applications are set in the *profile* module.

Table 5.2 Communication modes, protocols, and models for different messages

Message types	Communication mode	Protocols	OPNET models
SAVs	Publish/subscribe	Mapped to data link layer	*ethernet_station*
P&C status	Client/server	TCP/IP	*ethernet_workstation*
Trip signals	Publish/subscribe	Mapped to data link layer	*ethernet_workstation*
GOOSE	Publish/subscribe	Mapped to data link layer	*ethernet_workstation*
File transfer	Client/server	TCP/IP	*Server*

Table 5.3 Configuration of typical application in OPNET Modeler

Application names	Parameters of applications	
Videoconference (trip signals)	Incoming stream Inter-arrival time (seconds)	Exponential (0.002)
	Outcoming stream Inter-arrival time (seconds)	None
	Incoming stream Frame size (bytes)	Constant (204)
	Outcoming stream Frame size (bytes)	Constant (204)
	Symbolic destination name	Video Destination_trip
	Type of service	Reserved (7)
FTP (file transfer)	Command mix (get/total)	75%
	Inter-request time (seconds)	Exponential (0.001)
	File size (bytes)	Constant (1000)
	Symbolic server name	FTP Server

Particularly, the transmission of burst-type GOOSE messages is implemented by elaborately configuring the parameters of *duration* and *repeatability of supported application* in the *profile* module. Taking the trip signal transmission and FTP file transfer as examples, the values of parameters in OPNET Modeler are shown in Table 5.3.

5.4 Case Studies

Four cases are considered for evaluating the real-time performance of the presented SCN. Case I studies the configuration principle of VLAN and verifies the superiority of this scheme. Metrics of network utilization and packet delay in normal condition are investigated particularly. Moreover, Case II carries out a comparative analysis for VLAN-based SCN between normal and fault conditions. Case III compares the real-time performance of star network with ring network. Besides, since the ring topology has an inherent capability of "self-healing" with RSTP configured appropriately, the impacts of ring broken by RSTP on overall performance VLAN-based SCN are investigated in Sect. 5.4.4.

5.4.1 Case I: Evaluation of VLAN Scheme

The simulation model for star network of SCN in OPNET is shown in Fig. 5.8.

Particularly, the VLAN configuration schemes in simulation study are described as follows:

(1) There are totally 12 VLANs configured for the whole substation, and the IDs for each VLAN are shown in Table 5.4.
(2) Typically, VLAN 1 contains all the IEDs in substation except MU IEDs.
(3) Two VLANs are configured for each of the transformer bays, which correspond to the measurement and protection unit for high-voltage and low-voltage side, respectively, while feeder bays and bus section bay have only one VLAN for each.

A more detailed result of VLAN configuration for D2-1-type substation can be obtained in Table 5.4.

In order to implement the VLAN schemes in OPNET, the specific ports of each bay switch, which are connected to the central switch, are required to set to "trunk" type, while other ports are "not configured."

The impacts of VLAN schemes on SCN performances are studied in this case. Figure 5.9 shows the packet delay characteristics of SCN in normal condition. It can be concluded that the Ethernet delay for SCN comes to 162.7 μs when there is no VLAN configured. However, as shown in Fig. 5.9, with the deployment of

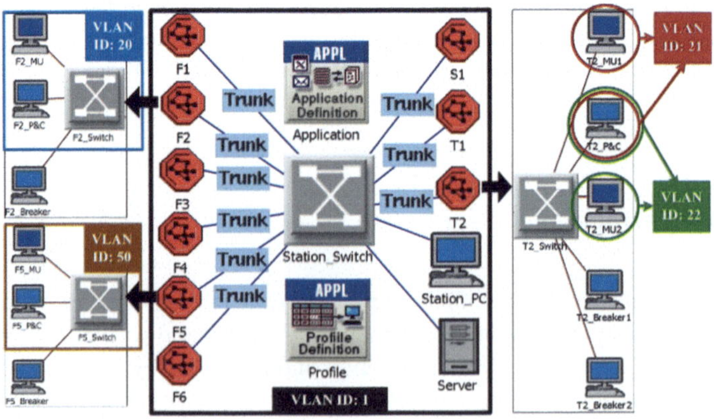

Fig. 5.8 Star network simulation model

Table 5.4 VLAN configuration results of substation

VLAN ID	Contained devices
1	All IEDs except MU IEDs
10	F1_MU IED, F1_P&C IED
20	F2_MU IED, F2_P&C IED
30	F3_MU IED, F3_P&C IED
40	F4_MU IED, F4_P&C IED
50	F5_MU IED, F5_P&C IED
60	F6_MU IED, F6_P&C IED
70	S_MU IED, S_P&C IED
11	T1_MU IED1, T1_P&C IED
12	T1_MU IED2, T1_P&C IED
21	T2_MU IED1, T2_P&C IED
22	T2_MU IED2, T2_P&C IED

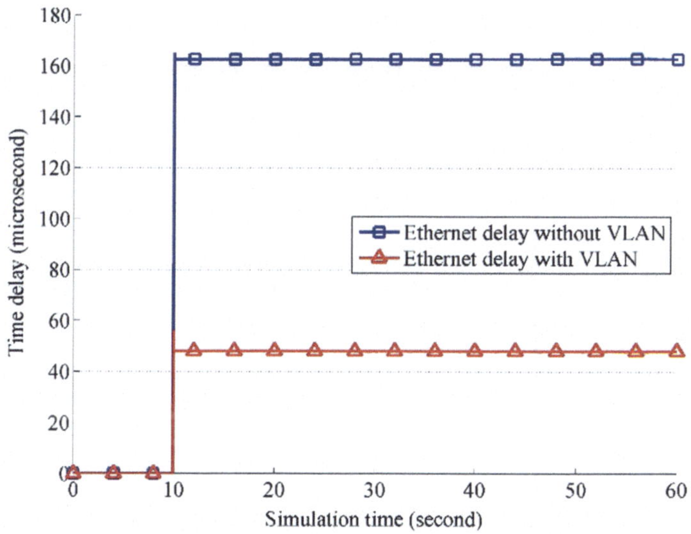

Fig. 5.9 Ethernet delay with/without VLAN in normal condition

VLAN schemes, the Ethernet delay has decreased to 48.09 μs, which indicate that the real-time feature of SCN has been improved greatly.

As the link utilization of network represents the usage levels of bandwidth, which to some extent reflects the remaining network resources available, the network utilization for various SCN links is investigated in this chapter.

Taking the T1 bay as an example, Table 5.5 shows the statistics of link utilization in the bay. Particularly, Fig. 5.10 shows the network utilization of output link between central switch and F1_swtich. It is obvious that the utilization for this link decreases

Table 5.5 Statics of link utilizations in T1 bay with/without VLAN

Link name	Link utilization without VLAN (%)	Link utilization with VLAN (%)
Switch → breaker IED1	51.74	0
Switch → breaker IED2	51.74	0
Switch → P&C IED	51.74	9.408
Switch → MU IED1	47.04	0
Switch → MU IED2	47.04	0
Central switch → switch	42.34	0
Switch → central switch	9.646	0.2424
MU IED1 → switch	4.704	4.704
MU IED2 → switch	4.704	4.704
Breaker IED1 → switch	0.0792	0.0792
Breaker IED2 → switch	0.0792	0.0792
P&C IED → switch	0.0792	0.0792

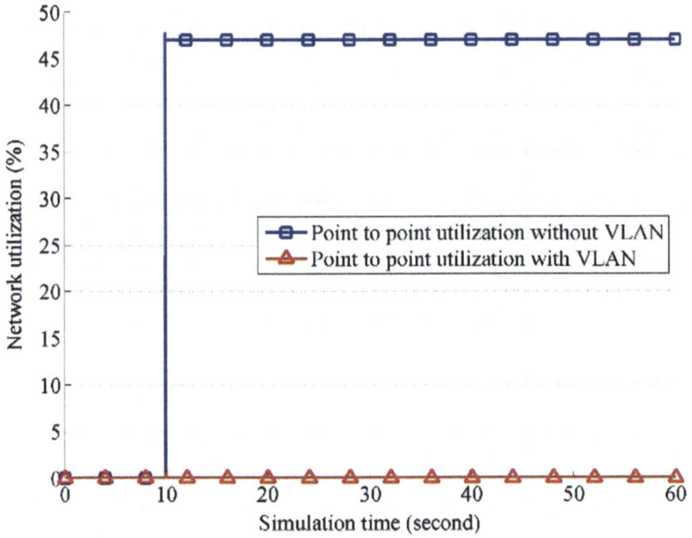

Fig. 5.10 Network utilization of output link between central switch and F1_swtich with/without VLAN in normal condition

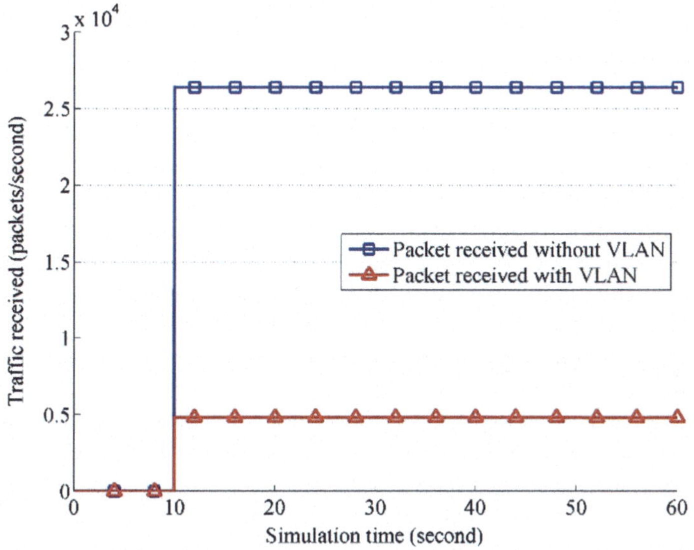

Fig. 5.11 Packet receiving rate of P&C IED in T1 bay with/without VLAN in normal condition

from 47.04% to zero with the application of VLAN. The reason is that the IEDs in F1 bay receive a large number of SAV packets from the MU IEDs in other bays when there is no VLAN configured. Consequently, the output link between central switch and F1_swtich is occupied partly by broadcasted SAV packets. However, with the deployment VLAN, SAVs are effectively limited within each bay; thus, there are not any packets on the output link between central switch and F1_swtich.

A similar conclusion can be obtained by Fig. 5.11, which shows the packet receiving rate of P&C IED1 in T1 bay under normal condition. It can be seen that when there is no VLAN, the packet receiving rate reaches as high as 26,400 packets/s, which is exactly the sum of packet sending rates for all the MU IEDs in substation. However, with the application of VLAN, the P&C IED in T1 bay receives the SAVs merely from the two MU IEDs within the bay, and the packet receiving rate is 4800 packets/s.

As stated above, the appropriate deployment of VLAN is able to limit a mass of SAV packets within each bay. Thereby, the data volumes among the bays can be reduced and the link utilizations between switches are lowered, which can reserve more bandwidth resources. Moreover, because of the superiority of VLAN, it is considered that the VLAN schemes presented in Table 5.4 have been well deployed for SCN in the subsequent case studies.

5.4.2 *Case II: Impacts of System Fault on Network Performance*

It has been described in Sect. 5.3 that the data amount transmitted in SCN increases greatly with the occurrence of system fault; thus, the impacts of a fault occurred in F1 bay on the network performance are studied particularly. As shown in Table 5.6, the statics of related link utilizations in normal and fault conditions are presented. As the link utilization varies greatly during fault period, the average values are calculated and listed in the third column of Table 5.6. It can be seen that the network utilizations of multiple links are increased, respectively, during fault condition.

Taking the communication link between central switch and F1_switch as an example, the change of network utilization for this link is shown in Fig. 5.12.

It can be seen that the network utilization of input link between central switch and F1_switch increases to 13.92% on average during fault period. The reason is that the communication loads between the two switches have increased greatly due to the transmission of trip signals and GOOSE messages with the fault occurred. While in normal condition the data transmitted between central switch and F1_switch contains only the cyclic meter values and status information generated by the IEDs in F1 bay, which have a relatively small data volume, thus the network utilization for this link is only 0.1616% without fault.

Figure 5.13 shows the variation of Ethernet delay for the overall network under normal/fault conditions. Particularly, the application of FTP file transfer is considered when the fault occurs. As shown in Fig. 5.13, at the normal operation, the Ethernet delay for data packets remains constantly with the value of 48.09 μs. However, with

Table 5.6 Statics of related link utilization in normal and fault conditions

Link name	Link utilization in normal condition (%)	Link utilization in fault condition (%)
Central switch → server	1.584	19.48
F1_switch → F1_P&C IED	4.704	17.18
F1_switch → central switch	0.1616	13.92
F1_Breaker IED → F1_switch	0.0792	11.58
S_switch → central switch	0.1616	13.02
S_Breaker IED → S_switch	0.0792	12.67
F1_P&C IED → F1_switch	0.0792	8.849
Central switch → F1_switch	0	7.601
Central switch → S_switch	0	1.737
S_switch → S_Breaker IED	0	1.691
F1_switch → F1_Breaker IED	0	1.75

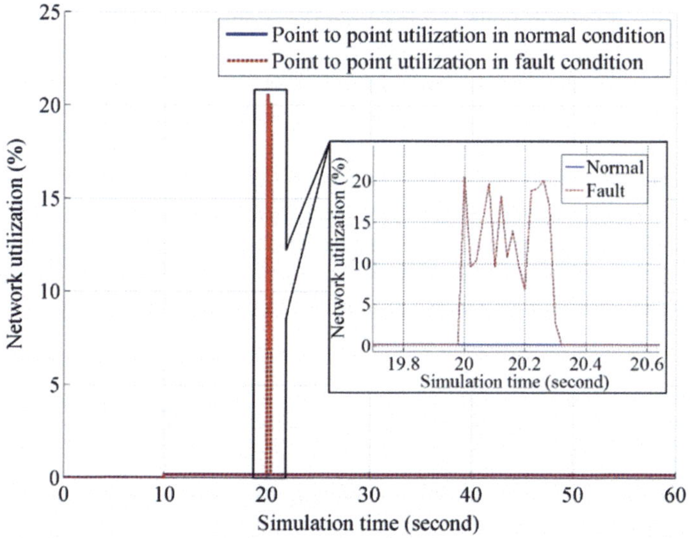

Fig. 5.12 Changes of network utilization of input link between central switch and F1_switch in normal/fault conditions

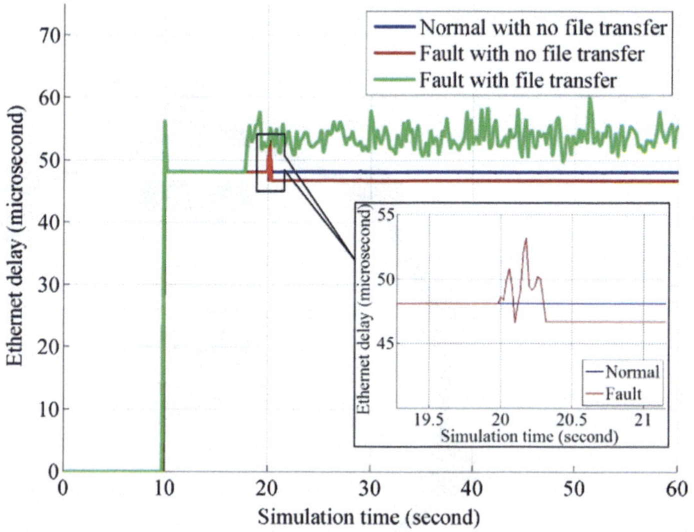

Fig. 5.13 Ethernet delay in normal/fault conditions considering file transfer

the occurrence of system fault, the Ethernet delay fluctuates and the average value during fault increases slightly to 49.63 μs.

Moreover, when the FTP file transfer is considered, the packet delay increases obviously according to Fig. 5.13. The mean Ethernet delay for SCN comes up to 53.59 μs. It can be concluded that the occurrence of system fault and file transfer will both cause the fluctuation and increase of packet delay because of the growing up of data amounts on the network. As the system fault generally lasts for several hundreds of milliseconds with the timely action of protection IEDs, it has little impacts on the overall real-time performance of SCN, but the increased data packets consume more bandwidth resources and may cause potential threats to network features.

5.4.3 Case III: Comparison of Ring and Star Topologies

The typical structure of SCN, however, could also be a ring topology [16, 18]. Therefore, a further study of characteristic for comparison of star and ring network has been carried out.

In order to avoid the broadcast storm, RSTP has been properly configured on ring network, as shown in Fig. 5.14. It can be seen that T1_switch is selected as the root bridge by RSTP and the port of F4_switch that connects F5_switch is blocked.

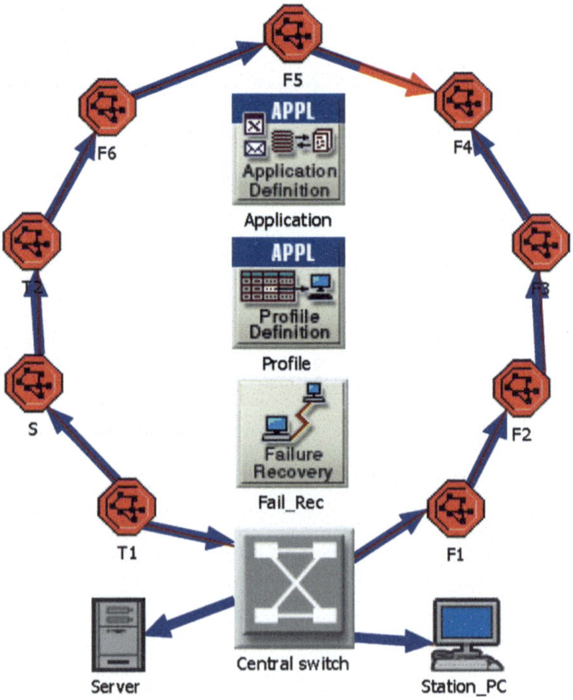

Fig. 5.14 Ring network simulation model

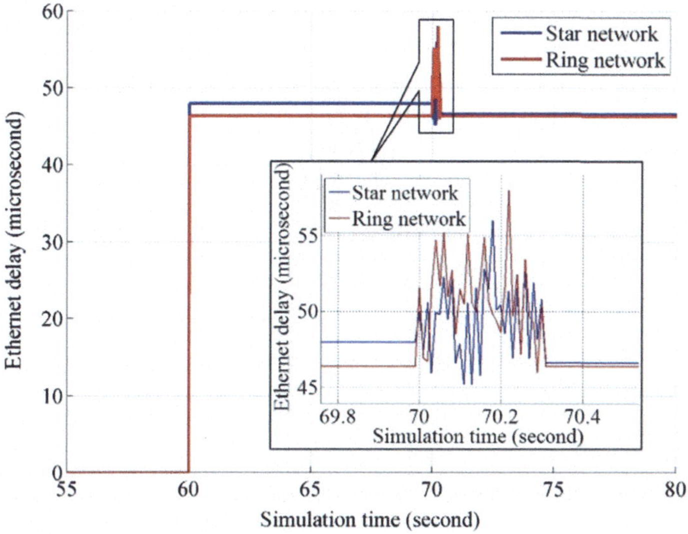

Fig. 5.15 Comparison of Ethernet delay for star and ring network

Moreover, since RSTP-based network needs several seconds to coverage during the initialization process, the start time of data generation in simulation as well as the occurrence of system fault is delayed accordingly.

According to Fig. 5.15, the Ethernet delay for ring network is a bit lower than star topology in normal condition, which are 46.38 μs and 47.97 μs, respectively. During system fault, the Ethernet delay for both structures increases and fluctuates simultaneously, as shown in Fig. 5.15. The average value of delay for ring structure grows up to 51.01 μs, which is a little bit higher than 49.60 μs of star network. Thus, it can be concluded that ring and star topologies of SCN are similar in the overall delay characteristics.

Figure 5.16 shows the variation of Ethernet delay for breaker IED in S bay. It is easy to know that the S_Breaker IED does not receive any packets in normal condition; thus, the Ethernet delay is zero. However, when a fault occurs, S_Breaker IED receives the trip signals from P&C IED1 in F1 bay, which causes a sudden increase in network load, and then the network delay of data packets grows accordingly. As shown in Fig. 5.16, during the fault time, the average Ethernet delay for S_Breaker IED in ring network increases to 57.81 μs while that of star network comes up to 70.32 μs. The real-time performance for ring topology is better than the star network with regard to time-critical trip messages, while both of the two structures can satisfy the requirement of IEC 61850 standard.

According to the above analysis, the star and the ring topologies are quite similar in overall real-time feature, while considering the time-critical messages the star network is a little poorer than the ring structure. Besides, star network is also poor in reliability that a single point of failure will cause the failure of data transmission. For example, if the link between F1 bay and central switch breaks, the communication

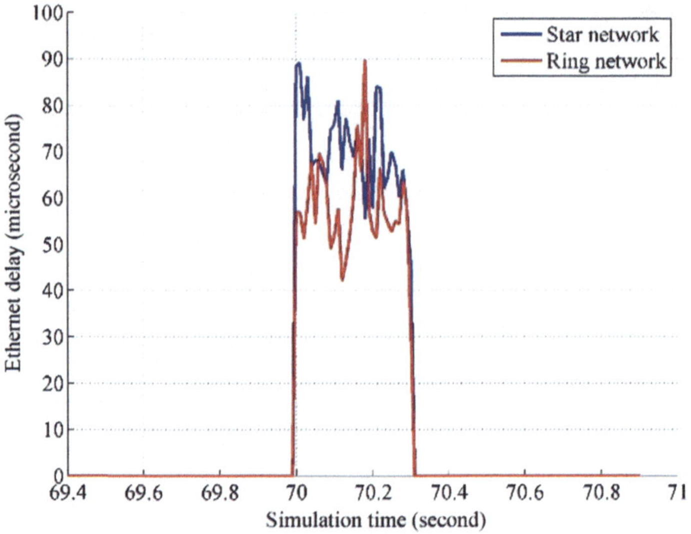

Fig. 5.16 Comparison of Ethernet delay for S_Breaker IED in star and ring network

from F1 bay to other bays and the station center will be stopped. However, ring network is simple in structure and low in cost. Moreover, ring topology has high reliability because of the "self-healing" ability. Therefore, it is preferred to apply ring structure for the proper design of substation network.

5.4.4 Case IV: Impacts of Ring Broken on Network Performance

According to the results in Case III, both of the star and ring networks are capable of transmitting messages within specified time limit. Although the ring topology has less time delay for some of data packets, the overall performance of the two structures is quite similar. Moreover, comparing to the star network, one of the main superiorities for ring topology is the intrinsic ability of "self-healing," which means the failure of one link or node in the ring would not necessarily cause the permanent breakdown of data transmission. Messages can be transmitted through another path in the reverse direction, and the data transmission is recovered after a certain time delay. As shown in Fig. 5.14, it is supposed that one of the communication links between different bays breaks down during simulation, and the impacts of ring broken by RSTP on the performance of SCN are evaluated particularly. The simulation time is extended further in order to observe the entire process of network response. Besides, as the system fault usually lasts shortly, it is not considered in this case. Supposing that the link between T1 and central switch breaks at 100 s, the simulation results are shown in Figs. 5.17, 5.18, and 5.19.

When the ring network is normal, the server in station level receives cyclic status messages from the P&C IEDs and breaker IEDs in all bays of substation, and the packet rate is 50 packets/s, according to Sect. 5.3. There are totally 20 P&C IEDs and breaker IEDs within the D2-1-type substation; thus, the throughput for the link from central switch to server with the ring is 1000 packets/s, as shown in Fig. 5.17. However, with the disconnection of the communication link between T1 and central switch, the throughput drops to 400 packets/s, which means that only 8 of the 20 IEDs maintain normal communication with the server. Further, we have found that the feeder bays of F1, F2, F3, and F4 still transmit data to server after the ring breaks because the broken link does not affect the original path of data transmission, while the rest of the bays fail to connect with the server once the link between T1 and central switch breaks down. Moreover, according to Fig. 5.17, it can be further concluded that the data transmission between central switch returns to normal at 141 s, which means that it takes the RSTP 41 s to recover the broken ring.

As mentioned above, with the broken link between T1 and central switch, RSTP needs 41 s to recover the network. However, each of the links may response asynchronously to the ring broken during the recovery process of SCN. Figure 5.18 shows the throughput for the communication links between different bays with ring broken by RSTP. Three different links are presented, which are $T1 \rightarrow S$, $S \rightarrow T2$, and $F4 \rightarrow F3$. It can be seen that once the link breaks, the throughput of $T1 \rightarrow S$ link increases to 150 packets/s. Nevertheless, the throughput of $S \rightarrow T2$ link changes from 0 to 250 packets/s at 118 s, which means this link does not respond to the network fault until 18 s after the ring broken. Moreover, at the time of 141 s, the throughput for the link $F4 \rightarrow F3$ increases from 100 to 700 packets/s, and the interrupted data transmission has been all recovered so far.

As each of the links in the ring network may go wrong occasionally, we investigated the network recovery time with different broken links. The results are shown

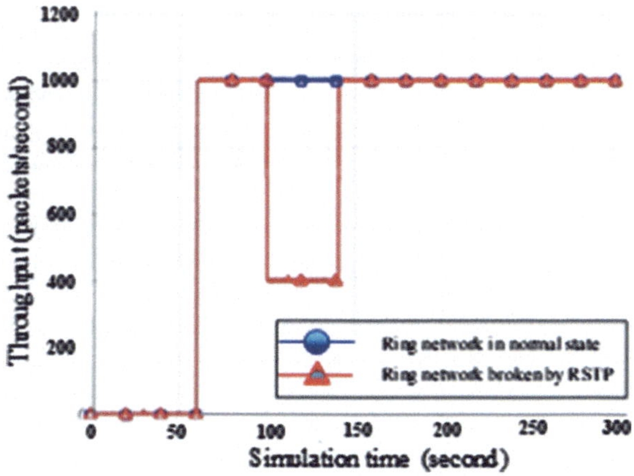

Fig. 5.17 Comparison of throughput for the link from central switch to server with the ring in normal and broken state

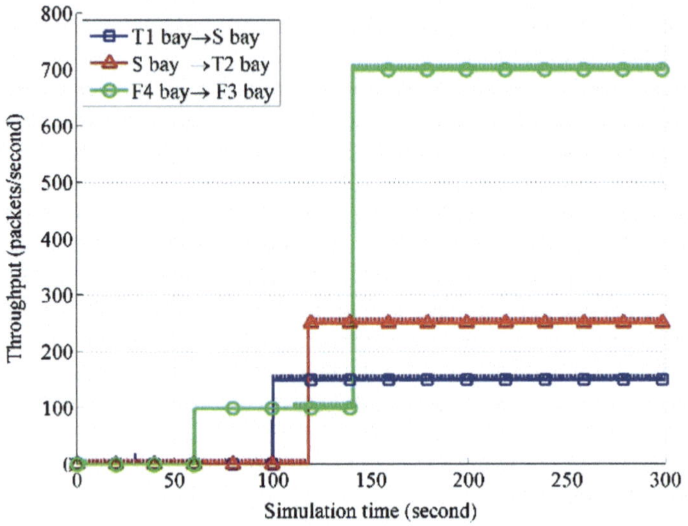

Fig. 5.18 Comparison of throughput for the communication links between different bays with ring broken by RSTP

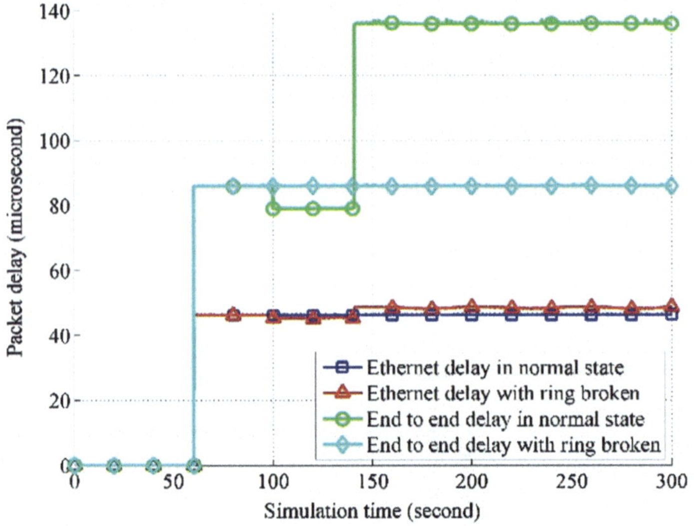

Fig. 5.19 Comparison of packet delay for the ring network in normal and broken state

in Fig. 5.20. It can be seen that the ring network needs the longest time of 41 s to recover when the link between T1 and central switch or S breaks down. However, if the broken link locates between F3 and F3 or F5, the network can restore immediately, which means the data transmission would not be interrupted. It can be further concluded that the network recovery time is related to the distance between the broken link and the RSTP root bridge. According to Fig. 5.14, we can see that the switch in T1 bay is selected as the root bridge by default. As shown in Fig. 5.20, it can be inferred that the network recovery time declines with the increase of distance from the link broken location to T1_switch. Particularly, the RSTP will block the switch port that has the maximum root path cost, such as the port of F4_switch which connects to F3_switch. Thus, the broken links nearby F4_switch will not affect the data transmission in ring network and the network recovery time is zero thereby.

In fact, the root bridge is generally determined during the initialization of RSTP, and the protocol selects the switch that has the minimum bridge ID as the root bridge by default. Therefore, once the root bridge is determined, the recovery time for ring network with different broken links can be evaluated accordingly. For the practical design and construction of ring SCN, it is preferred to improve the reliability of the links closer to root bridge, because the broken links may cause a longer disruption time of data transmission.

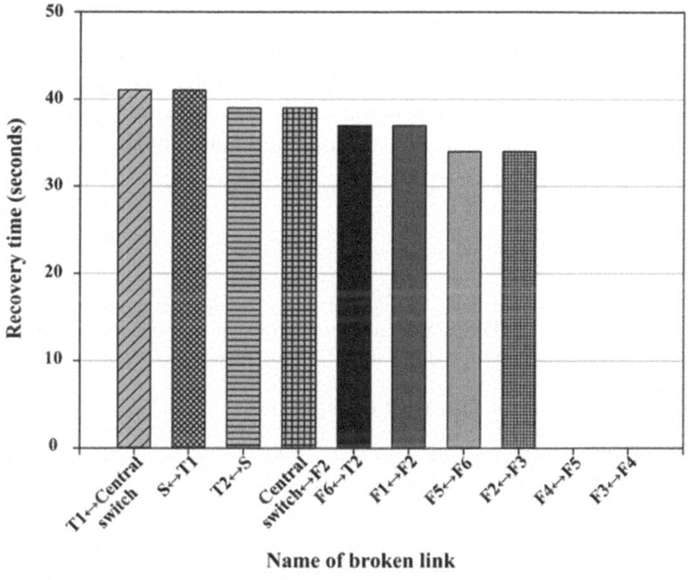

Fig. 5.20 Network recovery time of SCN with different links broken by RSTP

5.5 Summary

In this chapter, we propose three types of theoretical models for the data flow in SCN, which are cyclic data flow, stochastic data flow, and burst data flow. Based on these models, a quantitative analysis of typical data flow is carried out and the real-time performance for a VLAN-based substation is evaluated. It can be concluded that the proper configuration of VLAN schemes can reduce the data flow significantly by limiting the cyclic SAVs within the bay and thus lower the utilization of network links as well as the Ethernet delay of SCN. However, with the occurrence of system fault and FTP file transfer, the real-time performance of SCN is affected. The substantial growth of stochastic and burst messages during and after fault may consume more bandwidth resources, thus causing potential threats to network features.

Moreover, the comparison of performance for star and ring networks indicates that the ring structure is superior in real-time performance and reliability; thus, it is preferred to apply ring topology for the proper design of SCN. Particularly, when the ring structure is broken by the failure, the RSTP-based network is able to recover the data transmission. The network recovery time is in inverse proportion to the distance between the broken link and the RSTP root bridge. Therefore, the reliability of the links closer to root bridge should be enhanced especially, because the broken links may cause a longer disruption time of data transmission in ring network.

For the practical planning and building of SCN, it is beneficial for communication engineers to thoroughly evaluate the dynamic performance of SCN by LAN simulation based on proper data flow models. A desirable network can be achieved by considering the balance between its reliability and real-time performance of the network as well as the costs for its construction and maintenance.

References

1. Sidhu T, Gangadharan P (2005) Control and automation of power system substation using IEC 61850 Communication. In: Proceedings of IEEE Conference Control Application, Toronto, ON, Canada, pp 1331–1336
2. Zhang Z, Huang X, Keune B, Cao Y, Li Y (2015) Modeling and simulation of data flow for VLAN-based communication in substations. IEEE Syst J 11(4):1–12
3. IEC 61850 (2003) Communication networks and systems in substations, part 5: communication requirements for functions and device models, 1st edn
4. Zhang Z, Huang X, He J, Yang Y, Cao Y (2013) Self-adaption packet-loss-based sampled value estimation algorithm and its error analysis. Automat Elec Power Sys 37(4):85–91 (in Chinese)
5. IEC 61850 (2003) Communication networks and systems in substations, part 8-1: specific communication service mapping (SCSM)—Mapping to MMS, 1st edn
6. Zhang Z, Huang X, Cao Y, He J, Yang Y (2011) Comprehensive data flow analysis and communication network simulation for virtual local area network-based substation. Power Syst Tech 35(5):204–209 (in Chinese)
7. Taqqu MS, Willinger W, Sherman R (1997) Proof of a fundamental result in self-similar traffic modeling. IEEE Comput Commun Rev 27(5):5–23
8. Song S, Ng JKY, Tang B (2004) Some results on the self-similarity property in communication networks. IEEE Trans Commun 52(10):1636–1642

9. Willinger W, Taqqu M, Sherman R, Wilson D (1997) Self-similarity through high-variability: statistical analysis of ethernet LAN traffic at the source level. IEEE/ACM Trans Networking 5(1):71–86
10. Liebeherr J, Burchard A, Ciucu F (2012) Delay bounds in communication networks with heavy-tailed and self-similar traffic. IEEE Trans Inf Theory 58(2):1010–1024
11. Min G, Jin X (2013) Analytical Modelling and optimization of congestion control for prioritized multi-class self-similar traffic. IEEE Trans Commun 61(1):257–265
12. Adler R, Feldman R, Taqqu M (1998) A Practical guide to heavy tails: statistical techniques and applications. Birkbauser, Boston, pp 27–53
13. Zhang J, Konstantopoulos T (2005) Multiple-access interference processes are self-similar in multimedia CDMA cellular networks. IEEE Trans Inf Theory 51(3):1024–1038
14. Kim T, LeeD Lee S (2014) Hierarchical multilevel on/off source traffic modeling for a warship combat system. IEEE J Ocean Eng 39(2):226–242
15. Arnold B (1983) Pareto distributions. International Cooperative Publishing House, Fairland, Maryland
16. Sidhu T, Yin Y (2007) Modelling and simulation for performance evaluation of IEC 61850-based substation communication systems. IEEE Trans Power Del 23(3):1482–1489
17. Manassero G, Pellini E, Senger E, Nakagomi R (2013) IEC61850 based systems—functional testing and interoperability issues. IEEE Ind Informat 9(3):1436–1444
18. Thomas M, Ali I (2010) Reliable, Fast, and deterministic substation communication network architecture and its performance simulation. IEEE Trans Power Del 25(4):2364–2370

Chapter 6
Reliability Analysis of Cyber-Physical Systems in Substation

With the increasing interaction between physical devices and communication components, the substation based on the IEC 61850 standard is a type of cyber-physical system, especially various intelligent electronic devices (IEDs) and station bus, process bus applied in digital substation, which put forward high requests to real-time performance and reliability of substation [1, 2].

This chapter proposes a reliability analysis method for substations with a cyber-physical interface matrix (CPIM). This method calculates the influences from both the physical device failures and the communication devices failures. Two indices, probability of load curtailments and expected demand not supplied, are used in the reliability analysis. Given the simplified model of the practical substation based on the Chinese IEC 61850 standard, the results show that the substation system has a potential risk of cascading failure under the cyber-physical fusion trend, as the failure in cyber layer would increase the power loss of the whole system. The changing magnitude of expected demand not supplied increased significantly with increasing transmission delay rate of the process bus.

6.1 Interactions Between Cyber Layer and Physical Layer in Substation

Over the years, CPEPS has attracted considerable attention given their wide applications in grids, intelligent robot networks, embedded systems, and other fields. A typical CPEPS is capable of real-sensing, dynamic control, and information services [3–5]. Rather than cascading outages in power systems [6], smart cyber systems provide better monitoring, transferring, and controlling functions for the substation, but produce a trade-off, as the substation will experience more cyber-attacks. The supervisory control and data acquisition (SCADA) system of a nuclear plant has recently experienced a severe cyber-attack [7], so the study of cyber security has become a hot topic in smart grids. However, the interactions between cyber devices

Y. Cao et al., *Cyber-Physical Energy and Power Systems*,
https://doi.org/10.1007/978-981-15-0062-6_6

and physical devices in substations based on the IEC 61850 standard might create new failure scenarios to substations. Thus, it is important to address the reliability of the substation considering the interactions between the cyber layer and physical layer.

6.1.1 *Simplified Model of the Substation System*

Figure 6.1a shows a logical view of an example substation network architecture [8], commonly known as the substation based on the IEC 61850 standard automation model. IEC 61850 specifies how instantaneous sampled value (SV) measurements shall be transmitted over an Ethernet network by a merging unit (MU) or instrument transformer with an electronic interface. The IEC 61850 standard establishes a unified protocol for communication. Based on the standard, the main physical components are transmission lines, buses, circuit breakers, and main transformers. The cyber layer is divided into station level, bay level, and process level [9]. As a communication bus and a process bus can transmit and receive digital signals between the process level and bay level, they establish a communication connection between the protection unit, merging unit, and circuit breaker.

The circuit breaker, as the connecting and coupling component between the physical layer and cyber layer, plays the role of a controlling terminal. The components of a cyber-layer under IEC 61850 standard mainly include the process bus, merging unit, and physical component protection unit [10]. The protection unit includes the transmission line protection unit, transformer protection unit, and bus protection unit.

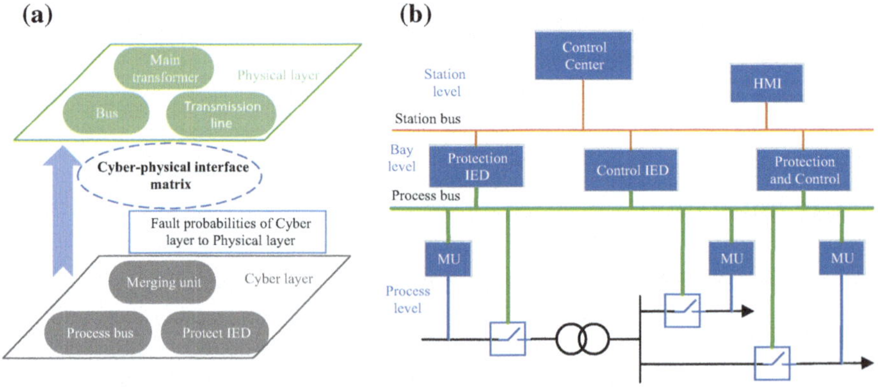

Fig. 6.1 **a** Components and **b** structure of cyber-physical substation system

6.1.2 Interaction Framework of the Cyber-Physical Substation

In Fig. 6.1, once one physical device breaks, the physical fault clearing process is the key factor for maintaining the correct functioning of the substation. The definition of fault clearing is when a physical component fails; the corresponding transformers or current transformers monitor the fault information, and then, send the analog signal to the merging unit (MU) [11]. The MU digitizes the information and sends the information to the protection intelligent electronic devices (IEDs) of the corresponding physical components. Protection IEDs generate the tripping signal through the protection algorithm. Finally, the process bus sends the signal to the circuit breaker for corresponding actions, thus limiting the scope of the failure of the physical components. This process is partially affected by the cyber components. If all the components in the process act normally, the fault clearing is successful, thus limiting the scope of the failure of the initial physical components. Otherwise, the fault clearing would fail and the scope of the failure may propagate to other physical components, and thus, the stable and secure operation of power networks would be further threatened [12].

As summary above, the reliability of cyber elements, such as MU, IEDs, and the process bus, is important to alert the primary equipment failure and help the substation continue working. Once some failures occur in the primary equipment in the substation, three types of scenarios occur during the physical fault clearing process, low-impact failure, local-impact failure, and wide-impact failure. Assuming a failure happened to the busbar, the three types of impacts are shown in Fig. 6.2. In this chapter, during the physical fault clearing process, if the related cyber devices work correctly, we call it working functionally and, otherwise, call it working malfunctioning.

The first type is the low-impact situation where no fault occurs in the cyber components (Fig. 6.2a). All the information from the primary equipment can be sent out; thus, the physical fault clearing process can work normally. For example, in Fig. 6.2a, the fault occurs in busbar and it does not spread elsewhere.

The second type is local-impact. Once some cyber components malfunction during the physical fault clearing process (excluding the process bus), the failures might spread to their surroundings, triggering them to malfunction, but the failure can be limited to the local scope by other functional cyber components. For example, in Fig. 6.2b, the initial fault also occurs in the busbar; the final fault spreads to the main transformer due to the MU failures.

The third type is wide-impact. The entire communication of the cyber-physical substation breaks down if the core of the communication components is damaged. For example, the process bus in the communication process plays the core role. Once it fails, all the information from the substation operation states would not be sent out. For example, in Fig. 6.2c, the initial fault still occurs in the busbar, and the whole system breaks due to the failure of the process bus.

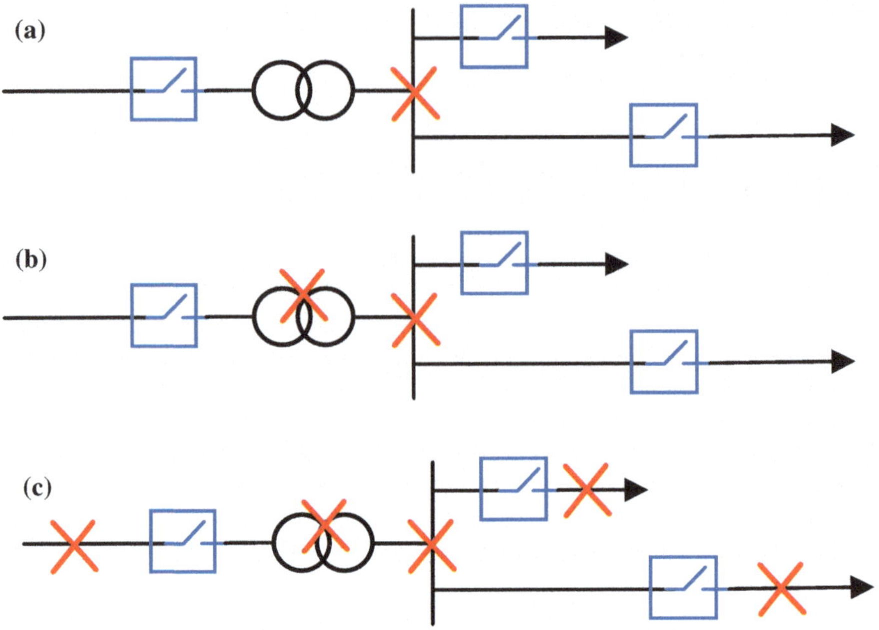

Fig. 6.2 Three types: **a** low-impact; **b** local-impact and **c** wide-impact of cascading failures in the substation

6.2 Model Quantifying the Interactions

Considering the three kinds of impact caused by cascading failures in substations, listed in Sect. 6.1.2, cascading failures chains can be described by a probabilistic model. To describe final cascading failure impact, we attempted to define the working states of the cyber components. A 0,1 sequence of related cyber components can reflect the final system state under different physical faults. For example, if 0 means functioning and 1 means malfunctioning, given the original failure in the substation, the working states of all related cyber components in the cascading failure chain can be obtained, and the impact of the cascading failure chain can be quantified as

$$\mathrm{CPM} = \begin{bmatrix} p_{1,1} & p_{1,2} & \cdots & p_{1,n} \\ p_{2,1} & p_{2,2} & \cdots & p_{2,n} \\ \vdots & \vdots & \ddots & \vdots \\ p_{m,1} & p_{m,2} & \cdots & p_{m,n} \end{bmatrix}_{\mathrm{m} \times \mathrm{n}} \tag{6.1}$$

where m is the number of the physical components, n is the amounts of cascading scenarios of each physical component; $p_{m,n}$ is the probability of causing the cascading scenario nth of the physical component mth; thus, the row vector $p_{m,i}$, $i \in [0, n]$, is the cascading scenario set of the physical component mth.

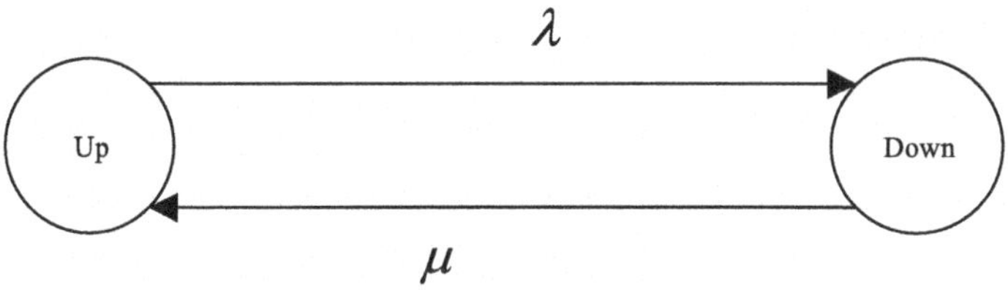

Fig. 6.3 Working states of cyber a component

Table 6.1 Data for cyber components

	Mean time to failure (years)	Failure rate λ (per year)	Mean repair time (h)	Repair rate μ (per year)
Protection IED	50	0.02	7.99998	1095.002
Merging unit	150	0.00667	7.99998	1095.002
Circuit breaker	100	0.01	7.99998	1095.002
Process bus	100	0.01	7.99998	1095.002

However, in practice, the cyber component working state is not actually 0 or 1. Thus, in the chapter, we modeled this as a two-state model, as shown in Fig. 6.3. The state of the cyber component is set to [0, 1], where 0 represents working functionally (down) and 1 represents working malfunctioning (up). In Fig. 6.3, λ denotes the failure rate of one individual component, and μ denotes the repair rate. The detailed data are given in Table 6.1.

The occurrence probability of a functionally working state p and unfunctional working state p' is calculated with Eqs. (6.2) and (6.3), respectively.

$$p = \frac{\mu}{\lambda + \mu} \tag{6.2}$$

$$p' = \frac{\lambda}{\lambda + \mu} \tag{6.3}$$

There are some delays in the communication process. The delay transmission of the process bus is denoted by probability η ($\eta = 0.3\%$ in the case study). Thus, Eqs. (6.2) and (6.3) can be updated as Eqs. (6.4) and (6.5) considering the delay, respectively.

$$p = \frac{\mu}{\lambda + \mu}(1 - \eta) \tag{6.4}$$

$$p' = 1 - p \tag{6.5}$$

Table 6.2 Working state probability of individual cyber components

Component	Functionally working state p	Unfunctional working state p'
Protection IED	0.999981735	0.000018265
Merging unit	0.999993912	0.000006088
Circuit breaker	0.999990868	0.000009132
Process bus	0.996990895	0.003009105

The functional working state and unfunctional working state probabilities of each cyber components are calculated as shown in Table 6.2. The functional working state probability of the process bus is smaller than that of the other components according to Eq. (6.4).

6.3 Reliability Analysis of the Cyber-Physical Substation

6.3.1 Indices of Cyber-Physical Substation Reliability

Probability of load curtailments (U_k) and expected demand not supplied (EDNS) were used to calculate the reliability of the cyber-physical substation, and they are displayed in Eqs. (6.6) and (6.7), respectively.

$$U_k = \frac{\sum_{i=1}^{N} T_{\text{dnik}}}{\sum_{i=1}^{N} \left(T_{\text{upik}} + T_{\text{dnik}}\right)} \tag{6.6}$$

where N is the number of the simulation, T_{dnik} is the duration of load k in ith curtailments, and T_{upik} is the duration of load k in the ith functionally working state.

$$\text{EDNS}_k = \sum_{i=1}^{N_k} P_{ik} L_k \tag{6.7}$$

where L_k is the average load not supplied of load point k during the simulation, P_{ik} is the probability of failure of substate i at load point k, and N_k is the total number of states or substates that cause load curtailment at load point k.

6.3.2 Reliability Simulation Method

The simulation was based on the sequential Monte Carlo method. Considering the cascading failures in the substation, the reliability simulation steps were as follows:

Step 1: Simulate time $t = 0$: Initialize both cyber layer and physical components.
Step 2: Randomly generate states of all physical components. The working state of each physical component is based on the exponential distribution:

$$T_i = -\frac{1}{\sigma_i} \ln U_i \tag{6.8}$$

where U_i of item i is within the interval [0, 1], which obeys uniform distribution. If the current working state of the item i is functional, σ_i is the failure rate of the physical component; otherwise, the current state is unfunctional, and σ_i is the repair rate of physical component. Finally, based on Eq. (6.8), we can find the $\min\{T_i\}$ and its corresponding component j. The working state of the physical component j will change at the next simulation time.
Step 3: The simulation time can be described as $t = t + 1$. Update the working states of all components.
Step 4: Calculate the cyber-physical interface matrix (CPIM) as shown in Sect. 6.2. Identify if a cascading failure happens according to Eq. (6.9). If so, then repeat step 3. Repeat this step until the failure no longer spreads. For component j, compare the value $p_{j,y}$ in the cyber-physical interface matrix (CPIM) with a random number P in the interval [0,1]. If P satisfies:

$$\sum_{y=0}^{s-1} p_{j,y} < P < \sum_{y=0}^{s} p_{j,y} \tag{6.9}$$

the sth scenario of the physical component j occurs
Step 5: Calculate the reliability indices.
Step 6: Repeat *Steps* 3–5 until the variance coefficient is less than the allowable value with

$$\beta = \frac{\sqrt{V(F)/NS}}{E(F)} \tag{6.10}$$

where $V(F)$ is the variance of the test function, NS is the number of simulation years, and $E(F)$ is the expected value of the function.

6.4 Case Studies

6.4.1 CPIM of the Reliability the Cyber-Physical Substation

A simplified model of a typical the substation based on the IEC 61850 standard in China is shown in Fig. 6.4, which is a 220/121/38.5 kV step-down substation. The

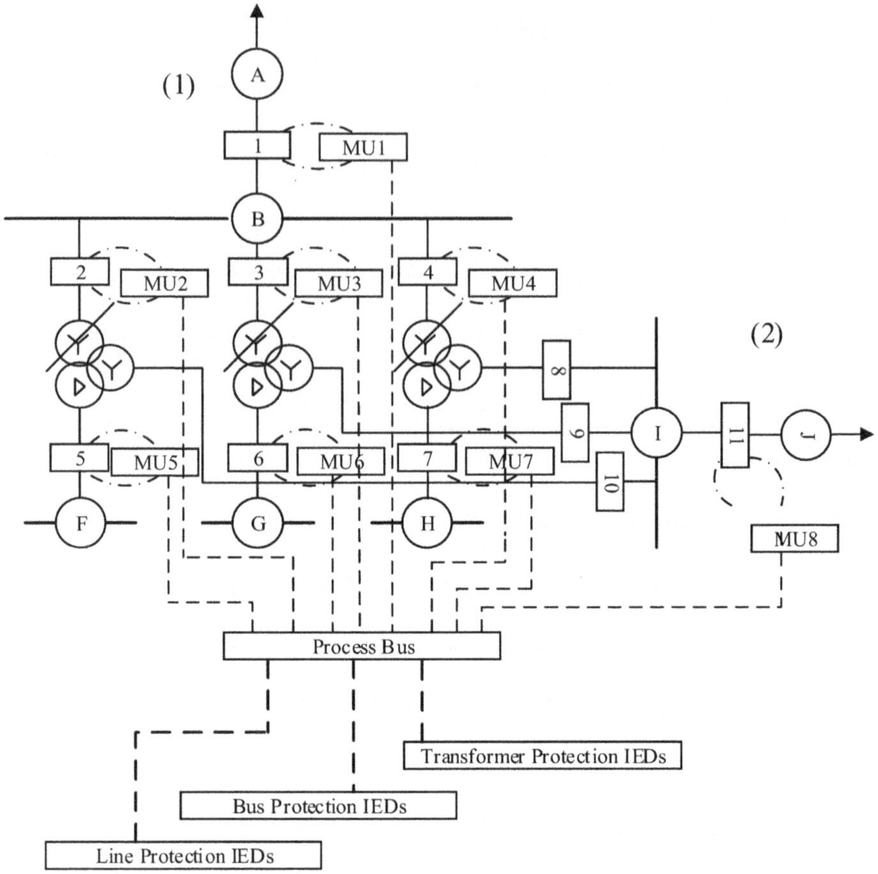

Fig. 6.4 Structure of a real IEC 61850 substation in China

annual average load of both load point 1 and load point 2 is 100 MW. The details for the primary devices of the substation are shown in Table 6.3.

In Fig. 6.4, there are 11 breakers, denoted as 1, 2, 3…; A and J stand for the transmission lines; C, D, E are main transformers; MU is the merging unit, and the number of MUs is 8, denoted as by MU1, MU2…; B, F, G, H, I are the buses. According to (6.1), the shape of the CPM of Fig. 6.4 is shown as (6.11). In (6.11),

Table 6.3 Equipment reliability data for the primary device

	Failure rate (per year)	Mean repair time (h)
Bus	0.002	13.0
Transformer	0.025	43.1
Transmission line	0.02	10.0

Table 6.4 Cyber-physical interface matrix (CPIM) of the line fault clearance at A

Cascading scenario	Effects scope	Probability
1	A	0.996957511
2	The entire system	0.000033384
3	AB	0.003009105

Table 6.5 CPIMJ of the line fault clearance at transmission line J

Cascading scenario	Effects scope	Probability
Low-impact	J	0.996957511
Wide-impact	The entire system	0.000033384
Local-impact	IJ	0.003009105

there are 10 physical devices, denoted as *A*, *B*...*J*; thus, the row number is $m = 10$; each row vector means the CPIM of a physical device. For example, the CPIM of the physical device *A* is denoted as $\mathrm{CPIM}_{A_{1\times a'}}$, where a is the number of cascading scenarios of *A*; similarly, the CPIM of the physical device B is denoted as $\mathrm{CPIM}_{B_{1\times b'}}$, where *b* is the number of cascading scenarios of *B*; the CPIM of the physical device J is denoted as $CPIM_{J_{1\times j}}$, where *j* is the number of cascading scenarios of *J*; thus, the number of columns of CPM is $a + b \cdots + j$. The CPIM of each physical device shows from Tables 6.4, 6.5, 6.6, 6.7, 6.8, 6.9, 6.10, 6.11, 6.12, and 6.13.

Table 6.6 CPIMB of the line fault clearance at bus B

Cascading scenario	Effect scope	Probability
Low-impact	B	0.996911991
Wide-impact	The entire system	0.003009105
Local-impact 1	(AB)/(BC)/(BD)/(BE)	0.000015173
Local-impact 2	(ABC)/(ABD)/(ABE)/(BCD)/(BCE)/(BDE)	1.38564396^{-10}
Local-impact 3	(ABCD)/(ABCE)/(ABDE)/(BCDE)	3.51492326^{-15}
Local-impact 4	ABCDE	1.82105929^{-5}

Table 6.7 CPIMF of the line fault clearance at bus F

Cascading scenario	Effects scope	Probability
Low-impact	F	0.996957511
Wide-impact	The entire system	0.000033384
Local-impact	FC	0.003009105

Table 6.8 CPIMG of the line fault clearance at bus G

Cascading scenario	Effects scope	Probability
Low-impact	G	0.996957511
Wide-impact	The entire system	0.000033384
Local-impact	GD	0.003009105

Table 6.9 CPIMH of the line fault clearance at bus H

Cascading scenario	Effects scope	Probability
Low-impact	H	0.996957511
Wide-impact	The entire system	0.000033384
Local-impact	HE	0.003009105

Table 6.10 CPIMI of the line fault clearance at bus I

Cascading scenario	Effect scope	Probability
Low-impact	I	0.996911991
Wide-impact	The entire system	0.003009105
Local-impact 1	(IJ)/(IC)/(ID)/(IE)	0.000015173
Local-impact 2	(IJC)/(IJD)/(IJE)/(ICD)/(ICE)/(IDE)	1.38564396^{-10}
Local-impact 3	(IJCD)/(IJCE)/(IJDE)/(ICDE)	3.51492326^{-15}
Local-impact 4	IBCDE	1.82105929^{-5}

Table 6.11 CPIMC of the line fault clearance at transformer C

Cascading scenario	Effect scope	Probability
Low-impact	C	0.996927164
Wide-impact	The entire system	0.003009105
Local-impact 1	(BC)/(CF)/(CI)	1.51734070^{-5}
Local-impact 2	(BCF)/(CIF)/(BCI)	2.30941925^{-10}
Local-impact 3	BCFI	1.82100387^{-5}

Table 6.12 CPIMD of the line fault clearance at transformer D

Cascading scenario	Effect scope	Probability
Low-impact	D	0.996927164
Wide-impact	The entire system	0.003009105
Local-impact 1	(DB)/(DG)/(DI)	1.51734070^{-5}
Local-impact 2	(DBG)/(DBI)/(DGI)	2.30941925-10
Local-impact 3	DBFI	1.82100387^{-5}

Table 6.13 CPIME of the line fault clearance at transformer E

Cascading scenario	Effect scope	Probability
Low-impact	E	0.996927164
Wide-impact	The entire system	0.003009105
Local-impact 1	(EB)/(EH)/EI)	1.51734070^{-5}
Local-impact 2	(BEF)/(EBI)/(EIH)	2.30941925^{-10}
Local-impact 3	BEHI	1.82100387^{-5}

$$\mathrm{CPM} = \begin{bmatrix} \mathrm{CPIM}_{A_{1\times a}} & \vdots & 0 & \vdots \ldots \vdots & 0 \\ 0 & \vdots & \mathrm{CPIM}_{B_{1\times b}} & \vdots \ldots \vdots & 0 \\ \vdots & & & & \\ 0 & \vdots & 0 & \vdots \ldots \vdots & \mathrm{CPIM}_{J_{1\times j}} \end{bmatrix}_{10\times(a+b+\cdots+j)} \tag{6.11}$$

Based on the CPIM method in Sect. 6.2, considering a failure clearing at line A, the CPIM is shown in Table 6.4. In this case, there are three kinds of cascading chains within the substation. Scenario 1: if all the related cyber devices are working functionally, the breaker can obtain the failure information, and then locate and clear the failure. The failure scope would be limited within A, which is the low-impact case mentioned in Sect. 6.1. In Table 6.4, the results show that when line fault clearance occurs at A, more than 99% failures are limited to within A. However, in extremely few cases, the failure scope would extend to the entire system due to the dysfunctional working of the process bus connected to A, which is the wide-impact case mentioned in Sect. 6.1. In Table 6.4, the probability of this occurrence is the smallest. With a small probability of 0.3%, among breaker 1, merging unit 1, and protection IED of A, more than one cyber device may be malfunctioning; thus, it leads to breaker 1 failure and then resulting in the failure of B. At this time, breakers 2, 3, and 4 can work functionally, thus limiting the failure scope to within A and B, which is the local-impact case mentioned in Sect. 6.1. Thus, based on Table 6.4, the number of cascading scenario is 3, and the CPIM = $[0.996957511, 0.003009105, 0.000033384]_{1\times3}$; it satisfied $\sum \mathrm{CPIMA} = 1$. Using the same method, the CPIM of the line fault clearance at transmission line J can be obtained as the CPIMJ = $[0.996957511, 0.003009105, 0.000033384]_{1\times3}$, results showing as Table 6.5.

Table 6.6 shows a similar analysis in the case of a failure clearing at bus B. In this case, consider all cyber devices are connected to B, such as merging units 1, 2, 3, and 4; breakers 1, 2, 3, and 4; and the process bus. The three kinds of cascading chains could occur within the substation: low-impact, wide-impact, and local impart. In Table 6.6, more than 99% of failures are limited to within B due to all the related cyber devices functioning properly. However, having a smaller probability 0.3%, the failure scope would extend to entitle system due to the dysfunctional working of the process bus. According to the different sizes of failure scopes caused by different related cyber devices, four kinds of local-impact may occur with minimal probability.

In Table 6.6, there are four types of local-impacts, denoted as local-impact 1, 2, 3, 4, and the number of cascading scenarios is 17. Local-impact 1: if one of the merging units or related breakers malfunctions, the failure effect scope would be limited to B and one of its connecting physical devices. The number of cascading scenarios belongs to local-impact 1 is 4. For example, either merging unit 2 or the breaker 2 is dysfunctional, while the others are functional, then the effect scope is limited to within B and C. Local-impact 2: if two of merging units or related breakers are dysfunctional, this case would limit the failure effect scope to B and two of its connecting physical devices. The number of cascading scenarios belongs to local-impact 2 is 6. For example, the effect scope ABE might result from the failure at breakers 1 and 4 and merging units 1 and 4. Similarly, if three (four) of the merging units or related breakers malfunction, this would limit the failure effect scope to B and three (four) of its connecting physical devices. The number of cascading scenarios belongs to local-impact 3 and local-impact 4 which are 4 and 1.

Thus, based on Table 6.6, the number of cascading scenarios is 17, and the CPIMB = $[0.996911991, 0.000015173, 0.000015173, 0.000015173, 0.000015173, 1.38564396^{-10}, 1.38564396^{-10}, 1.38564396^{-10}, 1.38564396^{-10}, 1.38564396^{-10}, 1.38564396^{-10}, 3.51492326^{-15}, 3.51492326^{-15}, 3.51492326^{-15}, 3.51492326^{-15}, 1.82105929^{-5}, 0.003009105]_{1\times17}$, it satisfied $\sum$CPIMB = 1. Using the same method, the CPIM of the line fault clearance at bus F, H, G, I can be obtained as follows: the CPIMF = $[0.996957511, 0.003009105, 0.000033384]_{1\times3}$, the CPIMG = $[0.996957511, 0.003009105, 0.000033384]_{1\times3}$, the CPIMH = $[0.996957511, 0.003009105, 0.000033384]_{1\times3}$, CPIMI = $[0.996911991, 0.000015173, 0.000015173, 0.000015173, 0.000015173, 1.38564396^{-10}, 1.38564396^{-10}, 1.38564396^{-10}, 1.38564396^{-10}, 1.38564396^{-10}, 1.38564396^{-10}, 3.51492326^{-15}, 3.51492326^{-15}, 3.51492326^{-15}, 3.51492326^{-15}, 1.82105929^{-5}, 0.003009105]_{1\times17}$. All results are shown in Tables 6.7, 6.8, 6.9, and 6.10.

Using the same analysis method, Table 6.11 shows the results under failure clearing at transformer C. The results' summary is similar to Table 6.6:

(1) More than 99% failures are low-impact, limited to within C.
(2) Within a smaller probability of 0.3%, the failure scope extends to the entire system due to the disfunctional working of the process bus, being a wide-impact.
(3) Local-impact is classified according to the failure number of the related cyber device, of which the occurrence has low probability. Thus, based on Table 6.11, the number of cascading scenarios is 9, and the CPIMC = $[0.996927164, 1.51734070^{-5}, 1.51734070^{-5}, 1.51734070^{-5}, 2.30941925^{-10}, 2.30941925^{-10}, 2.30941925^{-10}, 1.82100387^{-5}, 0.003009105]_{1\times9}$, satisfied $\sum$CPIMC = 1.

Using the same method, the CPIM of the line fault clearance at transformer D, E can be obtained as follows: CPIMD = $[0.996927164, 1.51734070^{-5}, 1.51734070^{-5}, 1.51734070^{-5}, 2.30941925^{-10}, 2.30941925^{-10}, 2.30941925^{-10}, 1.82100387^{-5}, 0.003009105]_{1\times9}$, CPIME = $[0.996927164, 1.51734070^{-5}, 1.51734070^{-5}, 1.51734070^{-5}, 2.30941925^{-10}, 2.30941925^{-10}, 2.30941925^{-10}, 1.82100387^{-5}, 0.003009105]_{1\times9}$. All results are shown in Tables 6.12 and 6.13.

Table 6.14 CPIM of the line Probability of load curtailments (PLC) comparison

Load point	Probability of load curtailments		Growth rate (Δ)%
	Without cyber layer (traditional simulation)	With cyber layer	
(1)	3.78466667^{-5}	3.95233333^{-5}	4.43
(2)	3.81300000^{-5}	3.92400000^{-5}	2.91
Entire system	7.59766667^{-5}	7.73433333^{-5}	1.80

Table 6.15 EDNS comparison

Load point	EDNS (MWh/year)		Growth rate (Δ)%
	Without cyber layer (traditional simulation)	With cyber layer	
(1)	3.785	4.236	11.93
(2)	3.813	4.208	10.36
Entire system	7.598	8.160	7.41

6.4.2 *Reliability Analysis Results*

Consider the reliability of load point 1, load point 2, and the entire system in Fig. 6.4. The probability of load curtailment (PLC) was calculated as shown in Table 6.14. A traditional simulation without considering the impact of the cyber layer and our method with integrated CPIM was carried out. As seen from the growth rate (Δ%), the probability of load curtailment slightly increased by 4.43% compared to without considering the influence of cyber layer. The improvement is not obvious compared with the traditional simulation, especially for the entire substation. The risk of cascading failure was low due to the high reliability of the cyber components.

Compared with the traditional simulation, the *EDNS* of entire substation has increase by 7.41%. Compared the results of Table 6.15 with Table 6.14, the failures in the cyber layer have more significant impacts on electricity unavailability than on the probability of load curtailment.

6.4.3 *Effects of Delay Rates*

The comparison of EDNS is shown in Table 6.15. The EDNS in load point 1 increased 11.93%.

Values from 0 to 0.005 were assumed to be the delay rates for all process buses. In practice, a delay rate may be prolonged due to electromagnetic interference which was be influenced by other factors. The quantitative relationship between simulation time and the ENDS is studied, and the results are shown in Fig. 6.5. The value of the

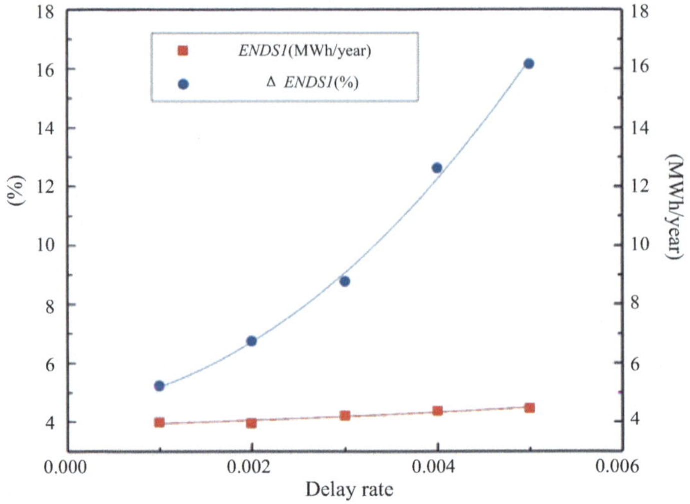

Fig. 6.5 EDNS and EDNS changing with delay rate at load point 1

system ENDS considerably increased and the growth rate of ENDS increased linearly with prolonged switching time. This illustrates that the delay rate of the process bus signifies the fault clearing. Advanced technologies for smart grids are important. Highly reliable control components and fast information transmission accelerate the process of cyber failure identification and physical fault clearing.

6.5 Summary

With the development of automation applications in substation system, the interdependency between the communication network and the primary equipment must be considered. This chapter extends the cyber-physical interface matrix (CPIM) methodology to reliability analysis. Two reliability indexes are presented, and the results of the case study verify that failures in the cyber layer increase the substation system's reliability, and the sensitivity analysis reveals that the process bus plays a key role in the reliability of the entire substation. Although the probability of time delay in information transmission is small, it is the critical factor leading to reliability changes in cyber-physical substations.

The proposed reliability assessment method can also be used to address the reliability problem faced by cyber-physical power systems. In such systems, for future study, it is needed that more detailed analysis on the interdependency between physical side and cyber layer.

References

1. Yi Y, Cao Y, Guo C, Liu B (2007) Design of IEC 61850 general gateway based on XML schema. Autom Electr Power Syst 31(2):60–64 (in Chinese)
2. Yi Y, Cao Y, Zhang J, Liu B, Xu L, Guo C (2008) A new centralized intelligent electrical device based on IEC 61850. Autom Electr Power Syst 32(12):36–40 (in Chinese)
3. Lee EA (2008) Cyber physical systems: design challenges. In: 2008 11th IEEE international symposium on object and component-oriented real-time distributed computing (ISORC), Orlando FL, USA, May 2008, pp 363–369
4. Anders GJ (1990) Probability concepts in electric power systems. Wiley, New York, NY
5. Cheng X, Lee W, Pan X (2017) Modernizing substation automation systems: adopting IEC standard 61850 for modeling and communication. IEEE Ind Appl Mag 23(1):42–49
6. Chen W, Jiang Q, Wang Z, Cao Y (2006) Risk assessment of cascading outages in power systems using fuzzy neural network. Lect Notes Comput Sci 3972:1422–1427
7. Bobbio A, Portinale L, Minichino M, Ciancamerla E (2001) Improving the analysis of dependable systems by mapping fault trees into Bayesian networks. Reliab Eng Syst Saf 71(3):249–260
8. Chen J, Thorp JS, Dobson I (2005) Cascading dynamics and mitigation assessment in power system disturbances via a hidden failure model. Int J Electr Power 27(4):318–326
9. Ferreira LC, Crossley P, Allan R (2001) The impact of functional integration on the reliability of substation protection and control systems. IEEE Trans Power Deliv 16(1):83–88
10. Cherdantseva Y, Burnap P, Blyth A, Stoddart K (2016) A review of cyber security risk assessment methods for SCADA systems. Comput Secur 56:1–27
11. Aghili SJ, Hoseinabadi HH (2017) Reliability evaluation of repairable systems using various fuzzy-based methods—a substation automation case study. Int J Electr Power 85:130–142
12. Li Y, Rehtanz C, Ruberg S, Luo L, Cao Y (2012) 6.Wide-area robust coordination approach of HVDC and FACTS controllers for damping multiple interarea oscillations. IEEE Trans Power Deliv 27(3):1096–1105

Chapter 7
Self-sustainable Community of Electricity Prosumers in Distribution System

Based on socio-technical system theory, the optimization of economic system can be achieved only by the society and technology being optimized simultaneously, which means the science, technology, and society are formed as a so closely linked giant system that we cannot consider each part separately. In distribution systems, as an example of the socio-technical system, there is a multitude of prosumers interact among themselves through a variety of ICT tools, in social networks, and with the electricity network.

In this chapter, we propose a new vision for the distribution system in which prosumers are encouraged by different balancing premiums to balance their electricity in a local community. Price-responsive generation and demand of an individual prosumer are affected by his/her attitudes and inherent characteristics. Further, the individual attitudes on benefit and comfort would evolve during social interactions. A load aggregator, on behalf of prosumers, participates in the community market run by a local electricity coordinator who takes responsibilities for operational securities and uses nodal prices to guide prosumers' physical behaviors. A regulator is assumed to design balancing premium schemes. A multi-agent-based simulation with a four-layered representation is employed to study features of the community and the best incentive strategies for the desired performance [1].

7.1 Self-sustainable Community for Electricity Prosumer

7.1.1 Characteristics of Self-sustainable Community for Electricity Prosumer

The power system is drastically changing with the smart devices, advanced information and communication technology (ICT), and active players [2, 3]. The fast development and vast deployment of the distributed generation urge exploitation, in a large scale, of cheap and clean renewable energies. Further, an increasing number

Y. Cao et al., *Cyber-Physical Energy and Power Systems*,
https://doi.org/10.1007/978-981-15-0062-6_7

of intelligent domestic appliances prompt the need for a smarter distribution system and a corresponding new energy vision.

For a better management of smart grids, many innovation concepts are proposed such as clustered-based power network for managing renewable energy resources and optimizing power flow exchanges. However, none of them addresses the impact from social behaviors and psychological characteristics of self-interested prosumers.

Actually, the emerging distribution system (EDS) is a socio-technical system [4, 5], characterized by the interactions between a technical infrastructure and social decisions based on different individual aspirations [6]. There are bidirectional interplays between social aspects that provide goals, expectations, "rules to play," etc., and technical structures that define physical functionalities, feasibilities, etc. The challenge of understanding such emerging systems is to design an accurate while flexible model where the new players [7, 8], resources [9], system rules, etc., as well as their interactions [10] and autonomous decisions [11, 12] are integrated to fulfill the whole electricity market and environment operating healthily.

From the perspective of power system, we consider the EDS as a whole, which not only focusing on one aspect as traditionally done (behaviors of consumers or network operation, or retailer market), but also providing a comprehensive framework in which the interplay of all the layers/players can be modeled. And prosumers are modeled as the elementary unit which produce and consume electricity continuously.

In the proposed new energy vision of self-sustainable prosumer-based community for the EDS, prosumers are incentivized to reach energy balance from near, green, economical sources, or self-organized solutions through corresponding BPs set by a regulator. The balancing levels include (1) self-supply, (2) self-organized trading at the same bus or in the same social circle, and (3) local community market (LCM).

The price and control signals are updated for every control time interval (CTI), depending on system requirements such as 15 min, 1-h, 1-day. A prosumer decides generation and demand for the next CTI. After self-supplying, he/she becomes a buyer/seller to balance the rest through self-organized trades (bilateral) or/and the LCM through a surrogate. The LEC, like a combination of DSO and a power exchange, is modeled to manage the network, including energy exchange with other communities through a higher voltage bus, solve congestions, and clear the LCM with nodal prices.

All prosumers have their own social time intervals (STI) which vary from individual to individual. After updating their attitudes on benefit and comfort, they will alter their generation and demand according to the new attitudes and the expectations on the price changes. Obviously, due to the differences in inherent stubbornness characteristics of everyone on price changes, the modification of the generation and demand is different for each prosumer even when they face the same price change. To assure the security of the network, the LEC acquires information on the self-organized trades and manages the system congestion along with the clearing of the LCM for each CTI.

7.2 Simulation Framework for Self-sustainable Prosumer-Based Energy Community

7.2.1 *Framework of Self-sustainable Community Simulation*

In this section, a four-layered framework integrates the social network, advanced ICT, power grids including the medium and low voltage levels and decision making of various players in different scales (Fig. 7.1).

Here, we can imagine the emerging active distribution system as a game, where different players are interacting and evolving in a defined environment to maintain the game, i.e., the whole electricity market and environment operating healthily. All having capabilities of decision making in their own domain according to specific goals, players in distribution system are divided into two classes. Major population under study is called micro-players, as they are driven and observed by the other

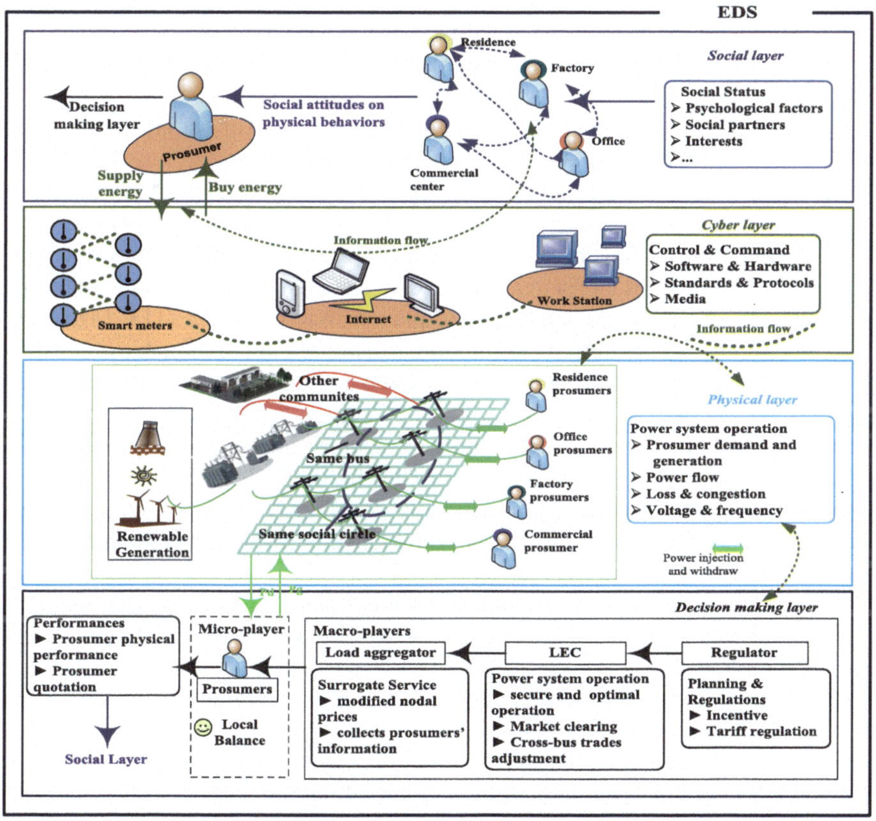

Fig. 7.1 Multilayer simulation framework for the EDS

players (macro-players) who set the rules, monitor trends, and coordinate among micro-players.

Examples of micro-players:

Prosumers (individual agents): persons, companies, institutions, and organizations that are connected to the distribution grid (MV and LV) in at least one point and that exchange energy (can be unidirectional as traditional customers and bidirectional as emerging prosumers) according to given power profiles continuously.

Retailers (market agents): companies whose core business is selling at the retail market of prosumer electricity. They may own some power capacity and purchase electricity on the wholesale market.

The other class, examples of macro-players:

Distribution system operators (DSO—system agent): organization that operates the distribution system with the goal to keep it feasible assuring some quality standards (continuity of supply, voltage distortion). They charge to prosumers and marketers fees for "transporting" power/energy to/from the prosumers according to some pre-defined quality standards. The quality standards are fixed by the regulator and improvement of worsening of the standards may result in prizes or penalties to the DSO.

Regulator (REG–regulating agent). Public body in charge of issuing the rules and exerting the control over the electricity (and more generally energy) markets.

The behaviors of a large amount of self-interested prosumers (an entity can both produce and consume energy), i.e. the micro-players pursuing individual economic benefit and comfort in the above mentioned environment, may exhibit self-organizational properties for (or against) the expectations from the society; therefore, their behaviors should be guided by macro-players in terms of rules [13, 14]. Regulation and policy-makers must, resorting to sophisticated decision making, consider the autonomous behaviors from all players and encourage active beneficial participations.

In the *social layer*, a prosumer randomly decides his/her social preference at each STI: benefit-driven or comfort-driven. Then he/she finds in his/her social circle a model prosumer of the same preference, who obtained the maximum benefit or comfort from last CTI. After that, he/she adjusts his/her attitudes toward those of the model prosumer.

The *ICT layer* provides a communication media for gathering and sending data regarding operational states, self-organized trades, the LCM results, and the current regulation regime.

In the *physical layer*, prosumers can inject and withdraw power into/from the network. The self-organized cross-bus trades are subject to the wheeling charge. Similarly, traditional generators and loads are also connected to this layer. The operational security of this layer is the responsibility of the LEC.

The *decision-making layer* is a virtual and private place for players to analyze and valid information from the ICT layer and then react based on individual objectives and constraints.

7.2.2 *Multi-agents Simulation Structure for Distribution Network*

In the proposed EDS, the balance is greatly encouraged, at the distribution level, within the community through the participation of the self-interested prosumers. It can mitigate the upstream grids in terms of managerial duties and power flow issues and consequently defer new investments. The only possible way to achieve this goal is to use effective signals (such as economic, cultural) to nudge the behavior of each prosumer toward the desired direction, which collectively decides the performance of the system. The control proposed in this chapter aligns the objectives of prosumers and macro-players so that the generated economic signals can be effectively used to induce the behaviors of prosumers.

The simulation structure (Fig. 7.2) contains the most typical characteristics and interactions of each player, abstracted from each layer in Fig. 7.1. Players thus can be divided into two general groups: micro-players, i.e., prosumers, and macro-players, including the load aggregator, the LEC and the regulator. Macro-players set rules, monitor the entire system, gather information, and issue-needed signals.

Prosumers share their information of latest comfort, benefit, and attitudes at each STI in the same social circle. At each CTI, prosumers analyze historical prices to predict possible variations of withdrawal and injection prices. Based on the newly

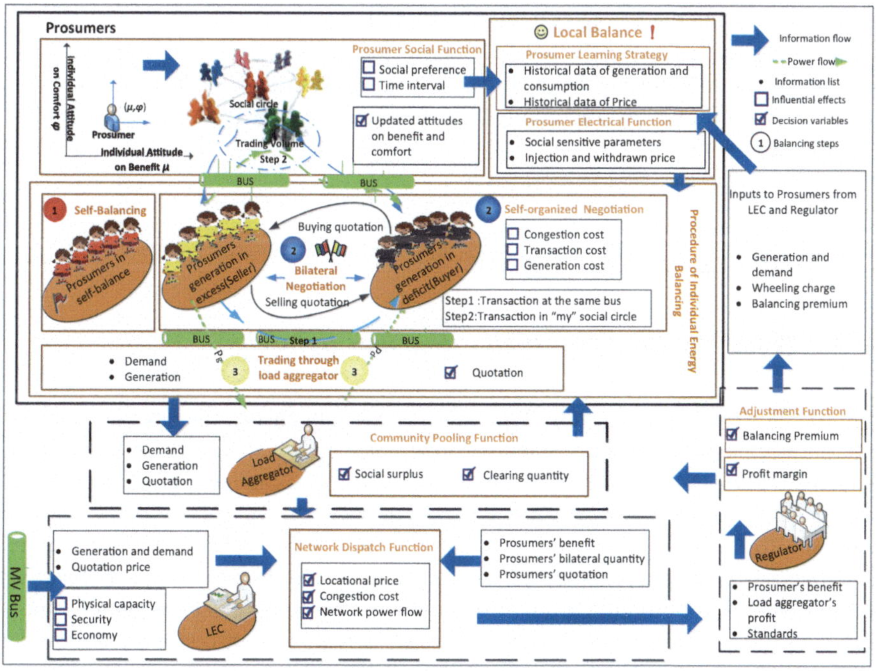

Fig. 7.2 Multilayer simulation framework for the EDS

updated social information and economic signals from macro-players, prosumers schedule their own generation and demand, considering generation and demand limits for the next CTI. After the self-supply, the role of a prosumer is defined as a seller (generation in excess) or a buyer (generation in deficit). During the participation of the trading in each CTI, a prosumer uses a profit-not-decreasing strategy to form the final price quotation. At each CTI, self-organized negotiation for the prices and quantities among prosumers within the same social circle or at the same bus considers the BPs and the wheeling charge as well.

The load aggregator provides services on behalf of prosumers, to participate in the LCM based on modified nodal prices, and collects the information of self-organized cross-bus trades and their corresponding adjustment bids to submit to the LEC. After receiving the clearing quantities and nodal prices from the LEC, the load aggregator recalculates the withdrawal and injection prices to obtain fee for the service. The clearing quantities and modified prices are then sent back to relevant prosumers.

The LEC runs the LCM using quotations from the load aggregator, small-sized traditional generators and loads in the same community, aiming at providing the most effective incentives to induce the behaviors of the market participants toward a desired outcome (such as maximizing the community social surplus, minimizing overall losses) while respecting to operational constraints. Energy exchanges with other communities are also considered as one of the LEC's tasks. Because the LEC cannot directly control prosumers' demand and generation, he/she uses the nodal prices as economic incentives to guide prosumers' decisions on their demand and generation, leading the whole system to a desired situation.

The regulator, as the only policy-making macro-player in our framework, is responsible for monitoring the operation of the market and the performances of the entire system. The "overall control" of the system can be performed by him/her through regulations.

7.3 Modeling for Micro-player

7.3.1 Modeling for Prosumer's Physical Behavior

To make decisions for the next CTI, prosumers not only change the price-responsive generation g_k^t and demand d_k^t, sensitive to μ_k^t and φ_k^t which can be influenced by social interactions, but also adjust quotations based on costs and profit-not-decreasing strategies.

By selecting different distributed equipment and considering the limit of equipment, the prosumer arranges their physical behavior according to the exact objective such as economic optimality, level, or sustainability objective at any moment. Therefore, the prosumer can be equivalent to a load, a generator, or a storage device, as Fig. 7.3 shows.

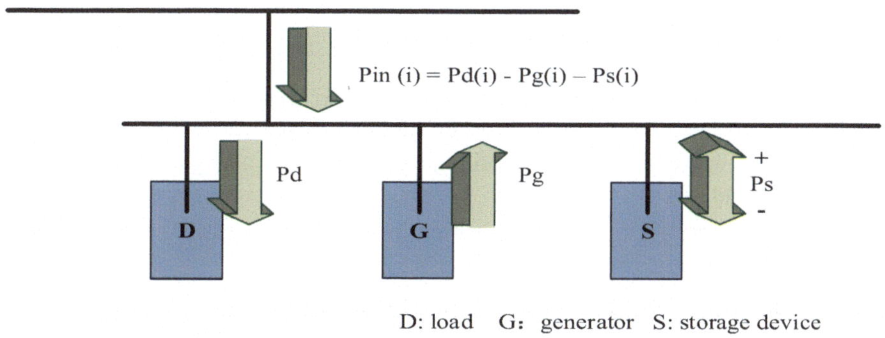

Fig. 7.3 Physical behavior of prosumer

Most common technical options and parameter variables for different prosumers can be listed as follows:

(1) Device options usually adopted by residential prosumers: photovoltaic, wind turbine, fuel cell, plug-in electric vehicle.
(2) Device options usually adopted by commercial and industrial prosumers: micro-turbines, hydro (depending on locations), flywheel (for storage of large capacity).
(3) Parameters for generation modules: levelized unit cost of electricity, max/min generation, efficiency of generation, lifetime.
(4) Parameters for storage modules: max/min storage capacity, charge/discharge rates, conversion efficiency, lifetime.

By analyzing historical data, the prosumers can be identified with three kinds of components mentioned above. And the behaviors of prosumer affect the efficiency of generation and storage, also indirectly affecting the surround environment, for example, the carbon emission. Thereby, in every market cycle, various devices can be simplified as power generation or consumption and the imbalanced energy of a prosumer can be determined by the generation g_k^t and demand d_k^t.

$$E_k^t = g_k^t - d_k^t \tag{7.1}$$

where E_k^t is the energy imbalance (kWh) of $\mathrm{k_{th}}$ set of prosumers at t; $g_{(\bullet)}^t$, $d_{(\bullet)}^t$ are the generation (kWh) and demand (kWh) of (•) at time interval t;

They are calculated as (7.2), (7.3), based on the reference generation and demand g_{nk}^t, d_{nk}^t which represent the probably available capacity and load profile for each CTI. Constant parameters $\vartheta_{k,g} \in [0,1]$ and $\vartheta_{k,d} \in [0,1]$ are introduced to model individual inherent demand-responsive and generation-responsive characteristics to price changes in (7.2), (7.3).

$$g_k^t = g_{nk}^t\left[1 - \mu_k^t(1 - \varphi_k^t)\bar{v}_k^t\vartheta_{k,g}\right] \tag{7.2}$$

$$d_k^t = d_{nk}^t\left[1 - \mu_k^t(1 - \varphi_k^t)\bar{\rho}_k^t\vartheta_{k,d}\right] \quad (7.3)$$

where μ_k^t, φ_k^t are the attitude of k on benefit and comfort at t; $\bar{\rho}_k^t$, $\bar{v}_k^t$ are the normalized withdrawal and injection price of k at t by dividing respective maximum absolute value of ρ^t, v^t which we define the price ρ^t is positive, v^t is negative while when electricity is sold out.

7.3.2 Modeling for Prosumer's Social Behavior

The evolution of prosumer k's individual attitudes on the benefit and comfort along time can be described by its trajectory on a normalized two-dimensional attitude space (μ, φ) $(\mu, \varphi \in [0, 1])$. The μ-axis signifies the attitude on economic benefit in terms of avoiding cost from consumption or maximizing earnings from generation. Likely, the φ-axis denotes the attitude on comfort in terms of the willingness to use appliances to satisfy his/her living standards [15].

Obviously, a prosumer concerns only the economic benefit if $\varphi_k^t = 0$: an increase in $\left|\bar{v}_k^t\right|$ or $\bar{\rho}_k^t$ increases the generation or decreases the demand according to (7.2) and (7.3). The magnitude of change then depends on his/her current attitude on benefit μ_k^t and his/her inherent characteristic ϑ_k.

For each social time interval (STI), a prosumer firstly decides his/her preference: benefit-driven (pursuing greater benefit) or comfort-driven (pursuing greater comfort). The benefit-driven (comfort-driven) prosumer then finds his/her model prosumer whose benefit (comfort) is the largest within his/her social circle. In fact, the absolute value of the benefit (comfort) is not strictly needed. Only a rank of the benefits (comfort) of the prosumers in his/her social circle is required to find the model. The rank can be generated through various methods, such as evaluation, conjecture. In this chapter, we simply use the absolute value of the benefit (comfort) calculated by (7.20), (7.21), and (7.22). By comparing his/her own $\left(\mu_k^t, \varphi_k^t\right)$ with those of the model prosumer $\left(\mu_m^t, \varphi_m^t\right)$, the prosumer's individual attitudes are updated to a new position $\left(\mu_k^t \pm \Delta\mu_k^t, \varphi_k^t \pm \Delta\varphi_k^t\right)$.

$$\Delta\mu_k^t = |m|\left(\mu_m^t - \mu_k^t\right)/\sqrt{\left(\mu_m^t - \mu_k^t\right)^2 + \left(\varphi_m^t - \varphi_k^t\right)^2} \quad (7.4)$$

$$\Delta\varphi_k^t = |m|\left(\varphi_m^t - \varphi_k^t\right)/\sqrt{\left(\mu_m^t - \mu_k^t\right)^2 + \left(\varphi_m^t - \varphi_k^t\right)^2} \quad (7.5)$$

where $|m|$ is the fixed magnitude in updating attitudes.

Figure 7.4 illustrates a simple case of 6 prosumers in the same social circle, with two model prosumers **Mb** (maximum benefit) and **Mc** (maximum comfort). Benefit-driven prosumers **A** and **B** move toward **Mb**. Comfort-driven prosumers **C** and **D** move toward **Mc**. Assuming prosumer **Mb** is still benefit-driven, he/she will not move; whereas **Mc** is benefit-driven now, thus he/she will move toward **Mb**.

Fig. 7.4 Procedure for updating prosumers' attitude

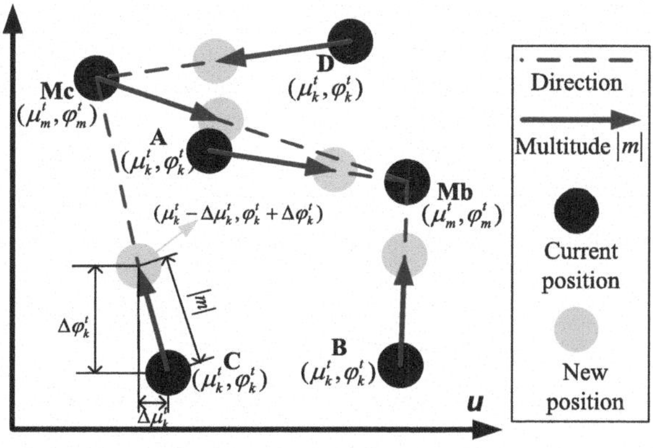

7.3.3 *Modeling for Prosumer's Self-organized Trade*

A buyer b asks offering prices and quantities firstly to all sellers at the same bus and then to a randomly selected number of sellers in his/her social circle. Thus, a list of preferential sellers is formed by (7.6), considering offering prices ε_s^t $\left(s \in \mathcal{K}_{b,s}^t\right)$, BP rewards ζ_b^t and unit wheeling charge ϖ^t. Similarly, a seller s forms a list of preferential buyers by (7.7). A transaction is successful only when buyer b and seller s are matched. The price and quantity of the transaction are agreed as in (7.8) and (7.9), respectively. To determine network-related costs, the information of cross-bus transaction quantities (7.10), gathered by the load aggregator, should be sent to the LEC.

$$\eta_b^t + \zeta_b^t - \varpi^t \geq \varepsilon_s^t \quad \forall b, \mathrm{s} \in \mathcal{K}_{b,s}^t \tag{7.6}$$

$$\varepsilon_s^t - \zeta_s^t + \varpi^t \leq \eta_b^t \quad \forall s, b \in \mathcal{K}_{s,b}^t \tag{7.7}$$

$$\psi_{s,b}^t = (\varepsilon_s^t + \eta_b^t)/2 \tag{7.8}$$

$$P_{s,b}^t = \min\left(E_s^t, \left|E_b^t\right|\right) \tag{7.9}$$

$$O_{s,b}^t = \begin{cases} P_{s,b}^t & s, b \notin i \\ 0 & s, b \in i \end{cases} \tag{7.10, 7.11}$$

where η_b^t, ε_s^t are the price quotation (¢/kWh) of b and s; ζ_b^t, ζ_s^t are the balancing premium (BP) reward (¢/kWh) to b and s for self-organized trade at t; ϖ^t are the unit wheeling charge (¢/kWh) at t; $\psi_{s,b}^t$ are the deal price (¢/kWh) for the self-organized trade of s and b at t; $P_{s,b}^t$ are the contracted energy (kWh) of s and b in self-organized trading at t; $O_{s,b}^t$ cross-bus contracted energy (kWh) of s and b in self-organized trading at t.

To decide the quotation for a transaction, a prosumer needs to consider the following two facts: For a buyer (7.12), (1) the bidding price should not be higher than the marginal utility (7.13), and (2) the expected benefit should not be lower than the benefit obtained at time $t-1$ (7.14).

For a seller (7.15), (1) the offering price should not be lower than the marginal cost (7.16), and (2) the expected benefit should not be lower than the benefit obtained at time $t-1$ (7.17).

$$\min \eta_b^t \tag{7.12}$$

$$\text{s.t.} \quad \eta_b^t \le f_b'(d_b^t) \tag{7.13}$$

$$f_b(d_b^t) - \eta_b^t |E_b^t| - f_b(g_b^t) \ge B_b^{t-1} \tag{7.14}$$

$$\max \varepsilon_s^t \tag{7.15}$$

$$\text{s.t.} \quad \varepsilon_s^t \ge f_s'(g_s^t) \tag{7.16}$$

$$\varepsilon_s^t E_s^t - f_s(g_s^t) + f_s(d_s^t) \ge B_s^{t-1} \tag{7.17}$$

where $f_{(\bullet)}(d)$, $f_{(\bullet)}^{'}(d)$ are the utility curve and marginal utility curve of $(\bullet)$, with $f_{(\bullet)}(d) = a_{1(\bullet)}d^2 + b_{1(\bullet)}d + c_{1(\bullet)}$, $a_{1(\bullet)} < 0$, while $f_{(\bullet)}(g)$, $f_{(\bullet)}^{'}(g)$ are the generation cost curve and marginal cost curve of $(\bullet)$, with $f_{(\bullet)}(g) = a_{(\bullet)}g^2 + b_{(\bullet)}g + c_{(\bullet)}$, $a_{(\bullet)} > 0$; B_k^t is the benefit (¢/h) and normalized comfort of k at t.

7.3.4 Modeling for Participation to Local Community Market

As we know the interaction between power grid and the dispatching data network will exert a tremendous impact on proper operation in dispatching center. We establish a more accurate interdependent network model, with several faults circumstances taken into consideration.

If the imbalance still exists after the previous two strategies, a prosumer can resort to the LCM through the load aggregator.

The price quotation is identical to (7.12) or (7.15), and the quantity R_k^t is calculated by (7.18) or (7.19).

$$R_k^t = \begin{cases} |E_k^t| - \sum\limits_{s \in \mathcal{K}_{b,s}^t} P_{s,b}^t & k \in \mathcal{K}_b^t \\ E_k^t - \sum\limits_{b \in \mathcal{K}_{s,b}^t} P_{s,b}^t & k \in \mathcal{K}_s^t \end{cases} \tag{7.18, 7.19}$$

After getting the cleared quantity and price (ρ_b^t, υ_s^t) from the load aggregator (7.23), the benefit B_k^t of prosumer k is computed by (7.20) or (7.21) depends on his/her role at time t, and the normalized comfort H_k^t is defined in (7.22).

$$B_k^t = \begin{cases} \left(\int_0^{G_k^t} f_k^{'}(d)\mathrm{d}d - \int_0^{G_k^t} f_k^{'}(g)\mathrm{d}g\right) + \left(\int_{G_k^t}^{G_k^t + \sum\limits_{s\in\mathcal{K}_{b,s}^t} P_{s,b}^t} (\zeta_b^t + f_k^{'}(d) - \psi_{s,b}^t - \varpi^t)\mathrm{d}d\right) \\ + \left(\int_{G_k^t + \sum\limits_{s\in\mathcal{K}_{b,s}^t} P_{s,b}^t}^{D_k^t} (f_k^{'}(d) - \rho_k^t)\mathrm{d}d\right) \quad k \in \mathcal{K}_b^t \\ \left(\int_0^{D_k^t} f_k^{'}(d)\mathrm{d}d - \int_0^{D_k^t} f_k^{'}(g)\mathrm{d}g\right) + \left(\int_{D_k^t}^{D_k^t + \sum\limits_{b\in\mathcal{K}_{s,b}^t} P_{s,b}^t} (\zeta_s^t + \psi_{s,b}^t - f_k^{'}(g) - \varpi^t)\mathrm{d}g\right) \\ + \left(\int_{D_k^t + \sum\limits_{b\in\mathcal{K}_{s,b}^t} P_{s,b}^t}^{G_k^t} (|\nu_k^t| - f_k^{'}(g))\mathrm{d}g\right) \quad k \in \mathcal{K}_s^t \end{cases} \quad (7.20, 7.21)$$

$$H_k^t = \varphi_k^t(1 - u_k^t) \quad (7.22)$$

where G_k^t, D_k^t are the final generation (kWh) and demand (kWh) of k at t.

7.4 Modeling for Macro-player

The load aggregator, on behalf of prosumers, participates the LCM and reports the self-organized cross-bus trades and corresponding adjustment bids to the LEC for scheduling the use of the network. After receiving the results of locational marginal price and quantity for each buyer and seller from the LEC, the load aggregator recalculates the price for each buyer (7.24) and seller (7.25). It is obvious that the load aggregator can gain extra profit through changing the nodal prices λ_i^t by the profit margin $r^t \in [0, 1]$ at t (7.23); thus, it must be regulated by the regulator through maximum allowed profit margin for the load aggregator $r_{\max}$ (7.26).

$$\max_{r^t} \sum_{\substack{d_b^{\prime\prime t} \\ b\in\mathcal{K}_b^t}} \rho_b^t d_b^{\prime\prime t} - \sum_{\substack{g_s^{\prime\prime t} \\ s\in\mathcal{K}_s^t}} |\nu_s^t| g_s^{\prime\prime t} \quad (7.23)$$

$$\text{s.t.} \quad \rho_b^t = (1 + r^t)\lambda_b^t \quad b \in \mathcal{K}_b^t \quad (7.24)$$

$$|\nu_s^t| = (1 - r^t)\lambda_s^t \quad s \in \mathcal{K}_s^t \quad (7.25)$$

$$0 \le r^t \le r_{\max} \tag{7.26}$$

The LEC is responsible for the operation of the network and the LCM which opens to traditional generators, loads, the load aggregator, and other communities. The optimization model (7.27) used by the LEC aims to

(1) maximize the social surplus through pooling traditional generators, residual loads, and excessive energy from prosumers;
(2) minimize the congestion cost [16]. The three items in (7.27) are the surplus from conventional loads and generators, surplus from prosumer sellers and buyers, and negative adjustment costs for self-organized trades, respectively. The interaction with other community (through a higher voltage bus) is simplified as an extra traditional market participant.

$$\begin{aligned}\max_{\substack{\Delta O^{t+}_{s,b},\Delta O^{t-}_{s,b}\\ g'^t_j, d'^t_z\\ g'^t_s, d'^t_b}} f &= \left(\sum_{z\in\mathcal{Z}} f_z(d'^t_z) - \sum_{j\in\mathcal{J}} f_j(g'^t_j)\right) + \left(\sum_{b\in\mathcal{K}^t_b} \eta_b d'^t_b - \sum_{s\in\mathcal{K}^t_s} \varepsilon_s g'^t_s\right)\\ &+ \left(\sum_{s\in\mathcal{K}^t_s}\sum_{b\in\mathcal{K}^t_b} \omega^{t-}_{s,b}\Delta O^{t-}_{s,b} - \sum_{s\in\mathcal{K}^t_s}\sum_{b\in\mathcal{K}^t_b} \omega^{t+}_{s,b}\Delta O^{t+}_{s,b}\right)\end{aligned} \tag{7.27}$$

$$\begin{aligned}\text{s.t.} \sum_{\substack{d'^t_b\\ b\in\mathcal{K}^t_b}} \left(\sum_{s\in\mathcal{K}^t_{b,s}} (O^t_{s,b} + \Delta O^{t+}_{s,b} - \Delta O^{t-}_{s,b}\right) + d'^t_b) + \sum_{\substack{d'^t_z\\ z\in\mathcal{Z}}} d'^t_z + F^t_{\text{loss}}\\ = \sum_{s\in\mathcal{K}^t_s} \left(\sum_{b\in\mathcal{K}^t_{s,b}} (O^t_{s,b} + \Delta O^{t+}_{s,b} - \Delta O^{t-}_{s,b}\right) + g'^t_s) + \sum_{\substack{g'^t_j\\ j\in\mathcal{J}}} g'^t_j\end{aligned} \tag{7.28}$$

$$F_{l,\min} \le F^t_l \le F_{l,\max} \tag{7.29}$$

$$V_{\min} \le V^t_i \le V_{\max} \tag{7.30}$$

$$0 \le g'^t_j \le g^t_{j\max} \tag{7.31}$$

$$0 \le d'^t_z \le d^t_{z\max} \tag{7.32}$$

$$d^t_{b\min} \le \sum_{s\in\mathcal{K}^t_{b,s}} (O^t_{s,b} + \Delta O^{t+}_{s,b} - \Delta O^{t-}_{s,b}) + d'^t_b \le d^t_{b\max} \tag{7.33}$$

$$g_{s\min}^{t} \leq \sum_{b \in \mathcal{K}_{s,b}^{t}} \left(O_{s,b}^{t} + \Delta O_{s,b}^{t+} - \Delta O_{s,b}^{t-}\right) + g_{s}^{\prime t} \leq g_{s\max}^{t} \tag{7.34}$$

$$0 \leq \Delta O_{s,b}^{t-} \leq O_{s,b}^{t} \tag{7.35}$$

$$\Delta O_{s,b}^{t+} \geq 0 \tag{7.36}$$

where $\omega_{s,b}^{t+}$, $\omega_{s,b}^{t-}$ are the adjustment bids (¢/kWh) for self-organized trade of s and b at t; F_{l}^{t}, $F_{l,\max}$, $F_{l,\min}$ are the line flow (MVA) at t and line flow limits (MVA) of l F_{loss}^{t} is the line losses of the system (kVA) at t; $g_{(\bullet),\max}^{\prime t}$, $g_{(\bullet),\min}^{\prime t}$ are the generation limits (kW) of ($\bullet$) at t; j, z are the traditional generator and load index.

Equation (7.28) is the system power balance. Equations (7.29) and (7.30) are the line flow limits and voltage constraints, respectively. Equations (7.31) and (7.32) are power constraints for traditional generators and loads. Equations (7.33) and (7.34) are the limits for prosumer's generation and demand. Equations (7.35) and (7.36) are the limits for the adjustment of self-organized cross-bus trades.

Present regulatory schemes are well entered in the tradition and history of the electricity industry. The fast-technological innovation opens a gate to a multitude of new opportunities in terms of markets, services and related operational strategies, which prompts the needs for a deep revision in electricity regulation and a dynamic update. Models and simulation tools like those proposed in this chapter can help this process, providing an "in vitro" environment for studying and testing possible alternative regulatory choices.

In this framework, the regulator mainly takes two responsibilities: (1) to evoke interests of prosumers to be balanced with the closest sources or in a self-organized way, through different BPs; (2) to monitor the performance of the system and regulate behaviors of other macro-players, especially the natural monopolists in the market. For example, the regulator sets a profit cap $r_{\max}$ for the load aggregator when he/she provides services to the prosumers.

7.5 Case Studies

In this section, we demonstrate the proposed framework of the self-sustainable community through a simulation platform, implemented in Python, based on multi-agents techniques.

Figure 7.5 presents in a pictorial way, the approach we propose for modeling the EDS as socio-technical systems. We show electricity infrastructure in terms of buses and lines (according to the IEEE 33-bus system), and we connect a given number of prosumers at each bus (totally 320 prosumers). For each prosumer, we show his/her position on the attitude space that represents his/her physiological attitude and his/her social circle that identifies his/her social interactions. The bidding/offering

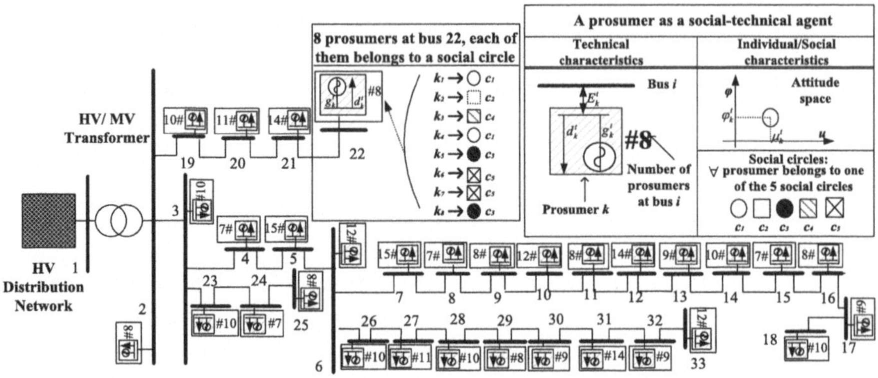

Fig. 7.5 Prosumers' community interacting with the distribution network (from IEEE 33-bus system)

curves for an equivalent outside community (located at bus 1) are assumed like internal traditional generators and loads. To minimize the impacts from non-prosumers' activities, we set $\Delta r_{\max} = 0.01$ and $\varpi^t = 0.5¢$. The CTI and STI are both assumed as 1 h. Prosumers' reference generation g_{nk} and demand d_{nk} are randomly initialized as $g_{\text{nk}} \in [5, 30]$, $d_{\text{nk}} \in [20, 30]$. The quadratic cost curves and utility curves are randomly initialized using the following parameters: $a \in [0.01, 0.25]$, $b = 20$, $c = 0$ and $a1 \in [-0.25, -0.01]$, $b1 = 32$, $c1 = 0$, respectively.

7.5.1 Impacts of Different Balancing Premium Schemes

It is important for policy decision makers to assess the impacts of regulations and the system responses to them ex-ante in order to devise the best BP schemes account for around 0–35% of the average price from the LCM. To focus on the comparison of different BPs, we set $\vartheta_{k,d} = 1$ and $\vartheta_{k,g} = 1$ to eliminate the influence from prosumers' demand-responsive and generation-responsive characteristics. We also compare the results with a reference case in which no social layer is modeled, i.e., the generation and load of each prosumer always equal to the reference values g_{nk} and d_{nk}. In addition, prosumers in the reference case participate directly to the LCM without trying to self-balance at first.

Figures 7.6 and 7.7 report the impacts of different BPs on prosumers' demand and generation, with zoom-in views for the quasi-stable parts. As the BP increases, prosumers increase their generation or demand for the self-organized trades to gain more rewards from the BP. As a consequence, both the demand and generation increase (cases of BP = 0¢, 3¢, 5¢). However, after reaching a certain level, the system becomes insensitive to the increase of the BP (case of BP = 8¢).

Figure 7.7 depicts the percentages of the sources from where the demands of prosumers are satisfied under different BPs. According to (7.6), (7.7), the BP can

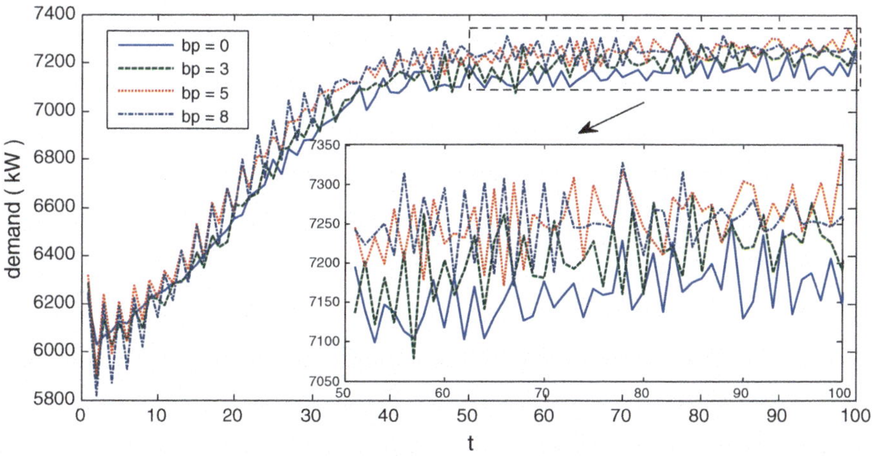

Fig. 7.6 Prosumer demand under different BPs

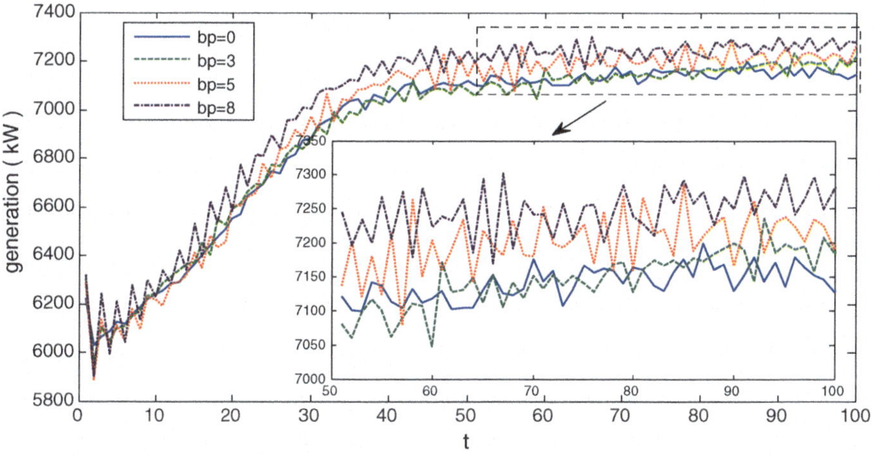

Fig. 7.7 Prosumer generation under different BPs

lower sellers' offering prices and increase buyers' bidding prices, then consequently encourages self-organized trades. Therefore, without the BP (BP = 0¢), the self-organized trading percentage is the lowest while the LCM percentage is the highest among the 4 BPs scenarios. After reaching a certain value, similar to Figs. 7.6 and 7.7, the BP becomes effectiveless in terms of encouraging self-organized trades of prosumers (Fig. 7.8b–d). It is because with the increase of the BP either demand or generation will firstly reach its maximum that the self-organized transactions hover at some level. After reaching this point, a prosumer seller (buyer) has to sell (buy) the imbalanced energy through the LCM. Yet, the bidding (offering) prices will be

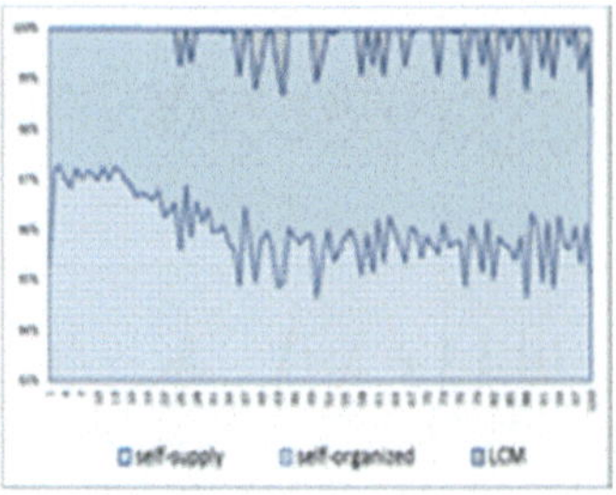

(a) BP= 0¢

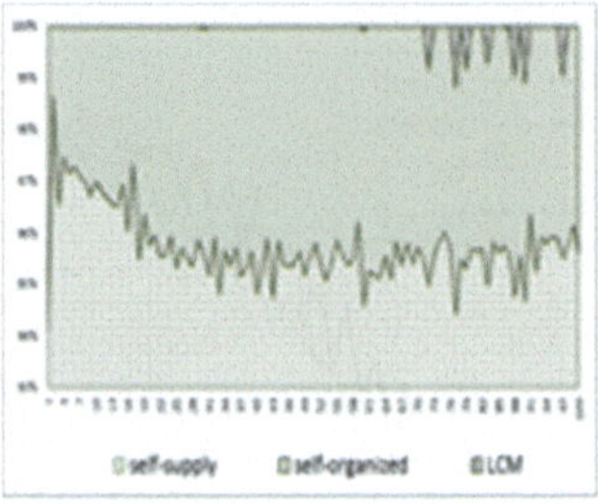

(b) BP= 3¢

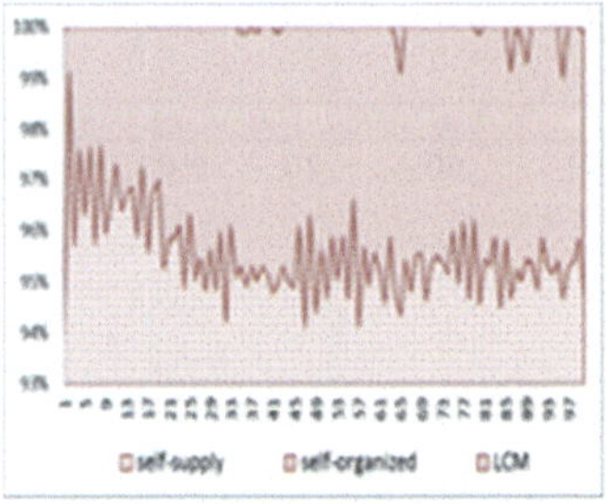

(c) BP= 5¢

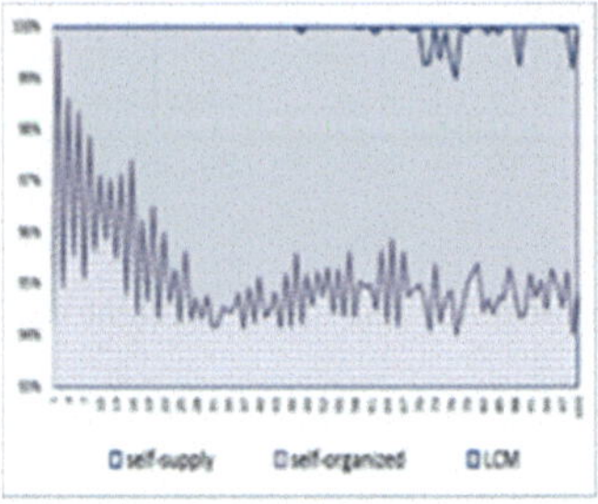

(d) BP= 8¢

Fig. 7.8 Shares in prosumers' demand (self-supply, self-organized, LCM) under different BPs

too low (high) to be cleared in the market. According to (7.12), (7.15), the prosumer will change the price quotation next time. In addition, since this prosumer failed in the market; thus, his/her benefit would not be the maximum among prosumers at his/her bus. Therefore, the social interactions will drive him/her toward the model prosumer in the attitude space.

The weighted market price is in a very narrow band from 22.35¢ to 23.55¢ and less than 1% of energy is traded through it (Fig. 7.8). Obviously, as the BP increases, the profit of prosumers rises even when the self-organized trade stops to grow.

Self-supplies and self-organized trades at the same bus meet most of the energy needs; therefore, power flow over the network decreases, which results in lower system losses (Fig. 7.9). As mentioned before, when BP $= 0$, both generation and load are the lowest in the simulation; therefore, the loss is also the lowest. Low cross-bus energy exchange indicates lower power flows over the network, which also implies less demand for infrastructure investment.

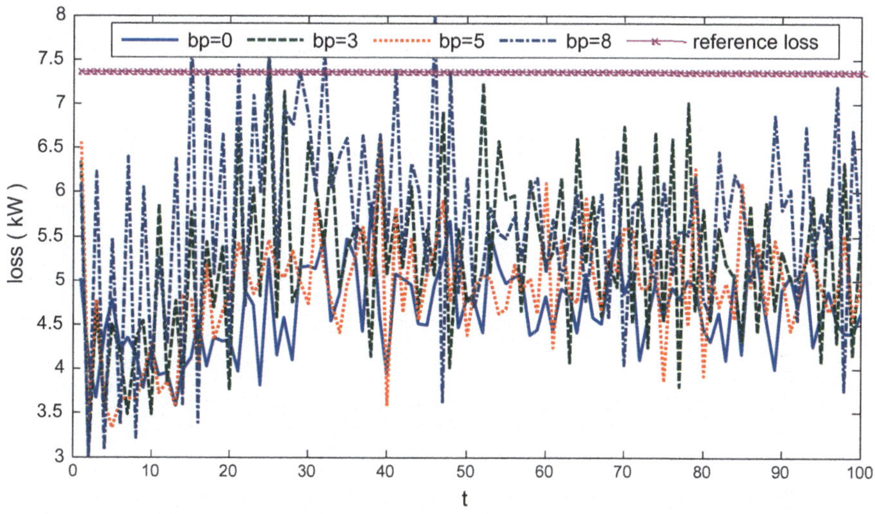

Fig. 7.9 IEEE 33-bus system loss under different BPs

7.5.2 *The Impacts of Prosumer's Inherent Characteristics*

In general, when the market price increases, prosumers lower the demand, and the opposite is true for the generation. However, each individual has his/her own psychological characteristics of price (7.2), (7.3). In order to study the impact of community members' stubbornness and noneconomic method to nudge prosumers' behaviors, we design three cases under the same BP (BP $= 5$¢) and assume $\vartheta_{k,g} = 1$, (1) 33% of members are sensitive to the price changes, whose $\vartheta_{k,d} = 1$ and the rest are with lower $\vartheta_{k,d} \in [0.3, 0.5]$; (2) 67% members with $\vartheta_{k,d} = 1$ and the rest with $\vartheta_{k,d} \in [0.3, 0.5]$; (3) all members with $\vartheta_{k,d} = 1$.

According to (7.3), the lower $\vartheta_{k,d}$ is, the less demand prosumers would decrease when withdrawal price increases; therefore, the more sensitive the prosumers are, the lower the demand is (Fig. 7.10). According to (7.6), with lower demand, buyers will increase the bidding prices, which is in favor of the self-organized trade (Fig. 7.11).

This behavior will then increase the profit of prosumers due to the BP rewards (Fig. 7.12). It is manifest that both the self-organized trades and average profit of prosumers grow when the number of responsive prosumers increases. In contrast, the demand and generation decrease when more price-responsive prosumers appear. This is a favorable outcome for the community as the total energy consumption decreases without sacrificing the economic profit of the community members.

As is shown in the simulation results, in order to maximize the advantages of the BP, policy decision makers need to cultivate common psychologies and culture of the community to be more responsive to the policies. Actions such as advertisement, propaganda, and transparent information can be useful to bring up the consensus of the community.

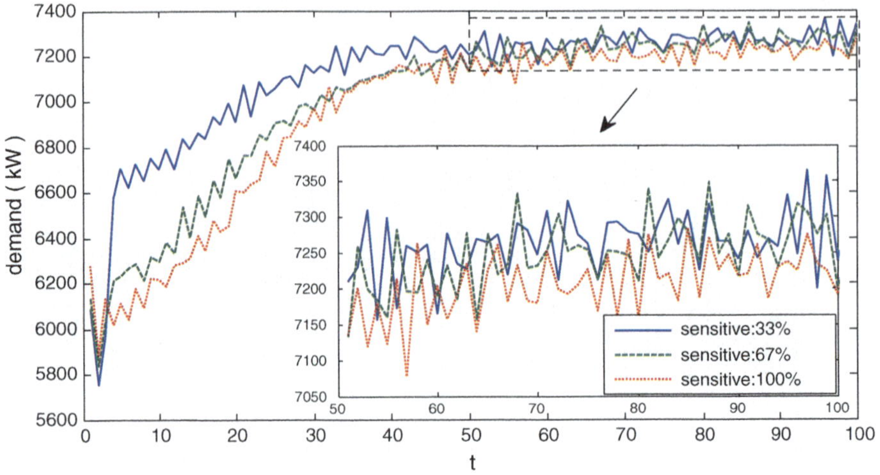

Fig. 7.10 Prosumer demand under different sensitivity cases

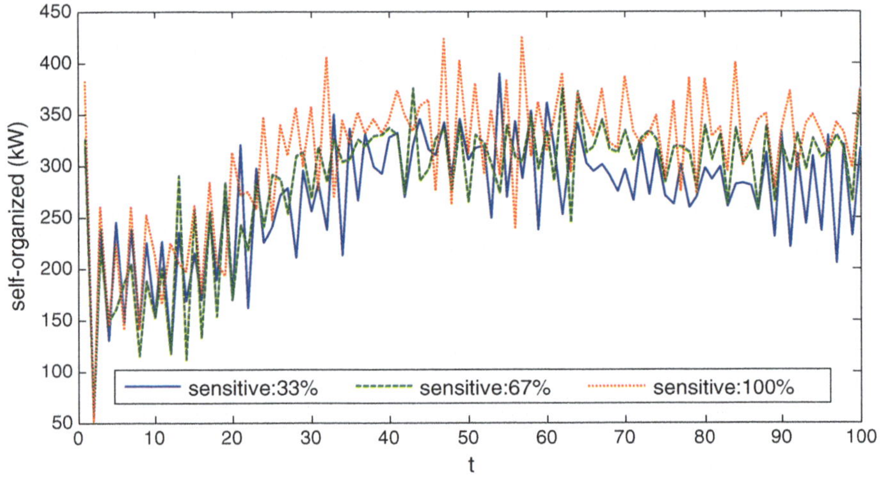

Fig. 7.11 Self-organized energy under different sensitivity cases

7.6 Summary

The emergence of prosumers changes the roles of the end users in the EDS from passively acquiring electricity to actively scheduling the demand and generation according to individual free will. It also makes the local exploration of renewable energy possible, and benefits both players and the system. However, new scenario brings extreme complexities to the study and control of the system. It prompts the

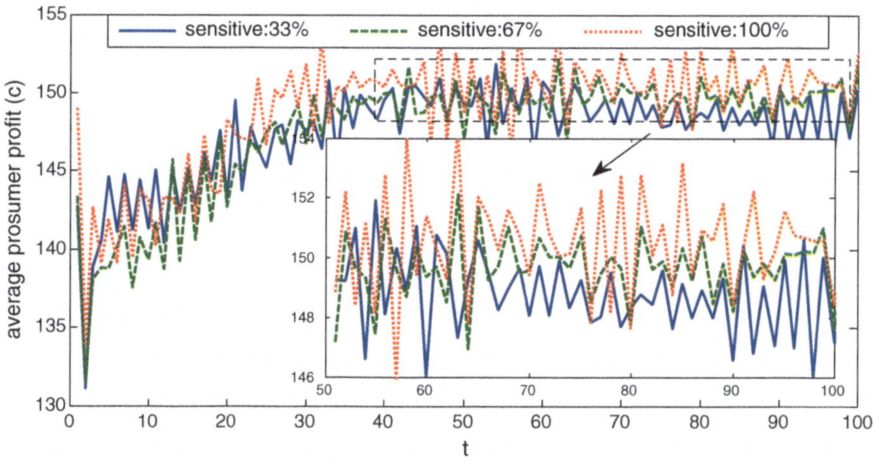

Fig. 7.12 Prosumer profit under different sensitivity cases

needs of new approaches and tools for understanding the EDS, especially for the policy decision makers. Our four-layered framework can study the interactive behaviors of the prosumers under the guidance from micro-players.

The emerging smart grid can open a new vision for the operation of power systems and make it possible to have a approach. Crucial in the merging framework are the social and technical interactions that affect the global performance and the needs to strive for macro-control of self-interested micro-players interacting at the social and network levels. The regulation can be made with external excitation in terms of both price and social values. Sometimes social values may be more effective than price signals. The results in this chapter show that to maximize the effectiveness of the policy, corresponding community culture should be carefully fostered. Our multilayered complex dynamic model is effective in capturing those interactions and in supporting policy decision making.

References

1. Cai Y, Huang T, Bompard EF (2017) Self-sustainable community of electricity prosumers in the emerging distribution system. IEEE Trans Smart Grid 8(5):2207–2216
2. Momoh JA (2009) Smart grid design for efficient and flexible power networks operation and control. In: 2009 PSCE'09. IEEE/PES, pp 1–8
3. Clastres C (2011) Smart grids: another step towards competition, energy security and climate change objectives. Energy Policy 39(9):5399–5408
4. Napoli R, Bompard E, Huang T (2009) Emergency coordination and decision making over interconnected power systems. In: 2009 IEEE international conference on systems, man and cybernetics (SMC), pp 3545–3551

5. Verbong G, Geels F (2007) The ongoing energy transition: lessons from a socio-technical, multi-level analysis of the Dutch electricity system (1960–2004). Energy Policy 35(2):1025–1037
6. Mohsenian-Rad A-H, Wong VW, Jatskevich J, Schober R (2010) Optimal and autonomous incentive-based energy consumption scheduling algorithm for smart grid. In: ISGT, pp 1–6
7. L-h Chang C (2013) The relationship among power types, political games, game players, and information system project outcomes—a multiple-case study. Int J Project Manage 31(1):57–67
8. Faruque A, Abdullah M (2014) RAMP: impact of rule based aggregator business model for residential microgrid of prosumers including distributed energy resources. In: Innovative smart grid technologies conference (ISGT), 2014. IEEE PES, pp 1–6
9. Nikolova S, Causevski A, Al-Salaymeh A (2013) Optimal operation of conventional power plants in power system with integrated renewable energy sources. Energy Convers Manage 65(65):697–703
10. Gungor VC, Sahin D, Kocak T, Ergut S, Buccella C, Cecati C, Hancke GP (2012) Smart grid and smart homes: key players and pilot projects. IEEE Ind Electron Mag 6(4):18–34
11. Samadi P, Mohsenian-Rad A-H, Schober R, Wong VW, Jatskevich J (2010) Optimal real-time pricing algorithm based on utility maximization for smart grid. In: 2010 first IEEE international conference on smart grid communications
12. Zhao P, Suryanarayanan S, Simões MG (2013) An energy management system for building structures using a multi-agent decision-making control methodology. IEEE Trans Ind Appl 49(1):322–330
13. Peik-Herfeh M, Seifi H, Sheikh-El-Eslami M (2013) Decision making of a virtual power plant under uncertainties for bidding in a day-ahead market using point estimate method. Int J Electr Power 44(1):88–98
14. Rathnayaka AD, Potdar VM, Dillon TS, Hussain OK, Chang E (2014) A methodology to find influential prosumers in prosumer community groups. IEEE Trans Ind Inf 10(1):706–713
15. Bompard EF, Han B (2013) Market-based control in emerging distribution system operation. IEEE Trans Power Deliv 28(4):2373–2382
16. Gawel E, Purkus A (2013) Promoting the market and system integration of renewable energies through premium schemes—a case study of the German market premium. Energy Policy 61:599–609

Chapter 8
Simplified Co-simulation Model for Investigating Impacts of Cyber-Contingency

In this chapter, a simplified co-simulation model is given for analyzing the interdependencies between energy and information flows and obtaining the quantitative relation between information features and power system operations. This co-simulation model features in low complexity and covering potential cyber contingencies. Moreover, the quantitative relation obtained by the model can provide requirements for the planning and operation of cyber-physical power systems.

8.1 Overview of Simulation Method

With integrations of information communication technologies (ICT) such as wide-area measurement systems (WAMS) and supervisory control and data acquisition (SCADA) systems, the power grids are evolving into a CPEPS, and CPEPS depends more on the capabilities of data communication and computation [1]. Due to tight cyber-physical couplings, cyber contingencies pose a significant impact on the physical process of CPEPS. For example, in [2, 3] the negative consequences of the cyber-induced deteriorated data are shown, and in [4–7], the adverse influence of the false-data-injection (FDI) and the denial-of-service-attacks (DoS) is elucidated.

Co-simulation is an efficient way to analyze and evaluate the impacts of cyber contingencies on power system applications including wide-area damping control (WADC), self-healing control of cyber-physical distribution network, and wide-area state estimation [8]. In traditional co-simulation methods, the power system and the cyber system are simulated by their dedicated simulation software. Especially, the cyber system is modeled from the perspective of the full process of cyber system operations, including protocol, exchanger, link, architecture, scheduling strategy, cyber domain entities, and so on. This method can simulate the dynamic process of both physical and cyber networks of CPEPS where the cyber system settings are deterministic.

As shown in Fig. 8.1, when investigating the impact of cyber contingencies, conventional methods firstly work on defining parameters of both power and cyber

Y. Cao et al., *Cyber-Physical Energy and Power Systems*,
https://doi.org/10.1007/978-981-15-0062-6_8

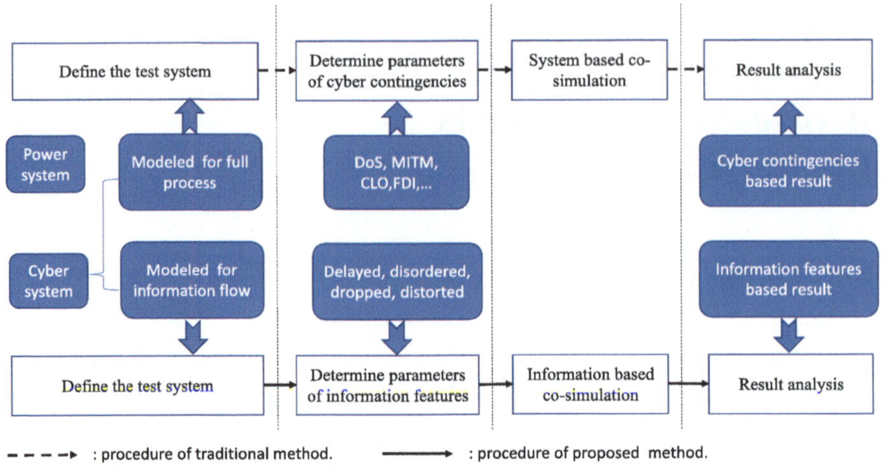

Fig. 8.1 Comparisons between conventional and proposed co-simulation models

systems, and then determine the property (e.g., type, occurrence place, operating mechanism, degree) of cyber contingencies. After that, each cyber contingency is realized in the cyber system simulation, according to the defined parameters. Eventually, through co-simulation between cyber and power systems, physical impacts of cyber contingencies can be investigated.

There are quite a few works on the conventional co-simulation method. For example, in [9], the PowerCyber testbed uses DIgSILENT PowerFactory, and RTDS to simulate the operation of a power system and Internet-scale event and attack generation environment to perform cyber attacks (malicious breaker trip and denial of service (DoS) with different attack throughput). In [10], the power system is modeled and simulated using RTDS and PSCAD. The simulation of the cyber system is implemented by Schweitzer Engineering Laboratories' devices, giga-transceiver network communication card-based simulated device, ns-3, and DeterLab. The proposed testbed analyzes impacts of communication line outage (CLO), DoS with different attack throughput, and man-in-the-middle (MITM) attacks on power systems under the specific cyber system settings. In [11], the co-simulation platform utilizes OpenDSS as the power system simulator and OPNET as the cyber network simulator, which provides a platform for testing different communication technologies. Based on DIgSILENT PowerFactory as the power system simulator together with OMNeT++ and INET Framework as the cyber system simulator, a co-simulation environment is presented in [12], which is capable of analyzing impacts of communication delay and failure. During the impact analysis of the communication delay, certain background traffic is added to achieve the desired delay. In [13], a co-simulation environment named Greenbench is presented for evaluating the impact of data-centric threats, which utilizes PSCAD and OMNeT++ for power and cyber system simulation, respectively. A cyber-physical testbed is introduced in [14] where the power system simulator RTDS is coupled with the cyber network emulator CORE to analyze the impact of cyber-attacks on power systems. Particularly, cyber-attacks are represented by the statue variation of physical equipment.

However, in the above works, there are two problems. The first one is that, when impacts of a given cyber contingency are going to be analyzed, the exact configuration of cyber systems is difficult to be implemented, which also decreases the accuracy of the cyber contingency assessment. The second one is that, when cyber system settings are deterministic, it is hard to cover the vast potential cyber contingencies.

However, for each operation cycle, CPEPS starts at physical systems and also ends in physical systems. The decision-making layer generates a decision based on measurements collected and transferred from physical systems. Similarly, the operation of physical systems is affected by the instructions issued from cyber systems, which can be regarded as the interaction between information and energy flows [15]. During the interaction, the information flow plays a significant role. Moreover, cyber contingencies affect the operation of CPEPS essentially by generating changes to the information flow, i.e., delaying, disordering, dropping, or distorting the data transmitted throughout cyber systems. Hence, it is feasible that the modeling of the cyber system is substituted by directly modeling the end-to-end information flow between all the starting/terminal nodes; Thus, the simulation can be implemented from a perspective of the information flow, rather than by simulating operations of the whole complicated cyber system.

8.2 Impacts of Cyber Contingencies

For security considerations, the cyber system of CPEPS usually utilizes the dedicated network that is physically isolated from public networks. However, it does not mean that cyber contingencies will not happen in cyber systems. For instance, attackers can steal into the computer room and hack computers by plug-in USB with malware. These hacked computers can be utilized to launch collective attacks. Besides, the increasing number of measurement devices, like PMUs, which are usually placed in the field without enough safety precautions, is possible to be hacked by adversaries for manipulating the measurement data. There is even a possibility that insiders (e.g., utility employees) directly implant attacks into the system [16].

The cyber contingencies can be merged and represented by changes in end-to-end information flows. Hence, in this section, it will be illustrated that the classifications of cyber-contingencies in different regions of CPEPS. Besides, the causes of cyber-contingencies and corresponding influences on the information flow are discussed, and end-to-end features adopted to describe the cyber-contingencies impact on the information flow are also presented.

8.2.1 Classification of Cyber Contingencies

The cyber system consists of sensing, communication, and decision-making layers. According to their influences, cyber contingencies can be categorized into three types as follows: (1) cyber contingencies in the sensing layer; (2) cyber contingencies in the communication layer; (3) cyber contingencies in the decision-making layer.

Cyber Contingencies in Sensing Layer: The sensing layer is the interface between physical and cyber systems. The layer converts analog quantities of the energy flow to digital quantities of the information flow and affects energy flows by delivering decisions made by the cyber system to actuators. According to the working status of devices in the sensing layer, cyber contingencies can be concluded into (1) disabled, (2) abnormal, (3) compromised equipment. For handicapped entities, due to random faults or malicious attacks, like the buffer overflow attacks, it becomes unable to send or receive data, which can be treated that all the corresponding data is dropped. For abnormal devices caused by the natural aging, temperature variation, or intentional causes of the magnetic and light attacks, the device data may suddenly deviate from the correct one. The compromised equipment means that the equipment is under the control of malicious attackers. Because the compromised equipment continues to generate and transfer false (distorted) data, which cannot be detected, it is more dangerous.

Cyber contingencies in communication layer: communication network layer consist of various end-to-end communication channels connecting the sensing layer with the decision-making layer. Based on the operating conditions of these communication channels, cyber contingencies can be summarized as: (1) interrupted, (2) re-routed, (3) congested, (4) incorrect transmission. For the channel interruption caused by random faults of communication devices or malicious attacks, all the data transferred by the channel can be defined as dropped. The channel re-routing always happens when the default nearest channel, which means the shortest transferring time, is interrupted, and there is more than one channel able to communicate the data. Since a longer channel is adopted, the consumed time of data transmission will be increased. The channel congestion brought out by a surge in the communication traffic will increase both transmission time and packet drop rate of data. The malicious attack is the most common element to cause incorrect channel transmission. For example, MITM sets up independent connections between the target endpoints and falsifies the data between them. Hence, the incorrect channel transmission can be represented by the distorted data communicated between the target endpoints.

Cyber Contingencies in Decision-making Layer: Cyber contingencies in the decision-making layer deteriorate operations of CPEPS by falsifying or interrupting the decision. For the example of the blackout happened in Ukraine, the attacker inserted the malware named BlackEnergy and KillDisk aiming at deleting the data in disk drives of the target computers to interrupt operations of the decision-making computer. Therefore, cyber contingencies in this layer can be depicted by the distorted decision data, which may trigger the maloperation of physical devices, and the dropped decision data, which prevents actuators from activating in time.

8.2.2 End-to-End Features of Cyber Contingencies

According to the above description, the impacts of cyber contingencies on information flows can be concluded into two kinds of properties: data and time. The former implies that random faults or malicious attacks may result in distorted data. The latter means that the consumed time related to processing, transmitting, and receiving processes between endpoints may cause the delayed, disordered, and dropped data. Hence, there are four kinds of adopted end-to-end information features, which are (1) delayed data, (2) disordered data, (3) dropped data, (4) distorted data.

Delayed Data: The time delay of data will cause an interval between the time when data is generated and the time when data arrives at the destination node. Define that the data is generated at the time t, and the corresponding time delay is t_r, $x(.)$ is the data at the destination node, and $x_s(.)$ is the data at the source node. The delayed data is as follows:

$$x(t + t_r) = x_s(t). \tag{8.1}$$

Disordered Data: Due to random time delays, the sequence of data arriving at the destination node may be disordered. Define contiguous data generated at the source node as $x_s(t)$ and $x_s(t + \Delta t)$, with time delay $t_r + 2 * \Delta t$ and t_r, respectively, where t is the simulation time step. The data arriving at the destination node is as follows:

$$x(t + t_r + 2 * \Delta t) = x_s(t). \tag{8.2}$$

$$x(t + t_r + \Delta T) = x_s(t + \Delta T). \tag{8.3}$$

which means the latter data $x_s(t + \Delta t)$ will arrive earlier than the data $x_s(t)$; thus, the data is disordered.

Dropped Data: The dropped data means the data is not received by the terminal node. It can be represented as:

$$\text{If } d = 1, \quad x(t) = \varnothing. \tag{8.4}$$

where d is the indicator of the dropped data. $d = 1$ or $d = 0$ means the data is dropped or not, respectively.

Distorted Data: The distorted data is:

$$x(t) = x_s(t) + \varepsilon, \tag{8.5}$$

where ε is the deviation from the correct data.

8.3 Information Flow-Based Co-simulation Model

There are four layers in the proposed co-simulation model: (1) power system layer; (2) sensing layer; (3) communication layer; (4) decision-making layer. Figure 8.2 shows the integrated simulations of the power system, sensing layer, communication layer, and decision-making layer.

8.3.1 Power, Decision-Making and Sensing and Communication Layers' Simulation

Power System and Decision-Making Layers' Simulation: The power system simulation is implemented in MATLAB for simulating dynamic characteristics of power system operations. The decision-making layer simulation is programmed by C++ language in Microsoft Visual Studio. It simulates operations of the designed power system application. Also, the computation time of the formed application is also

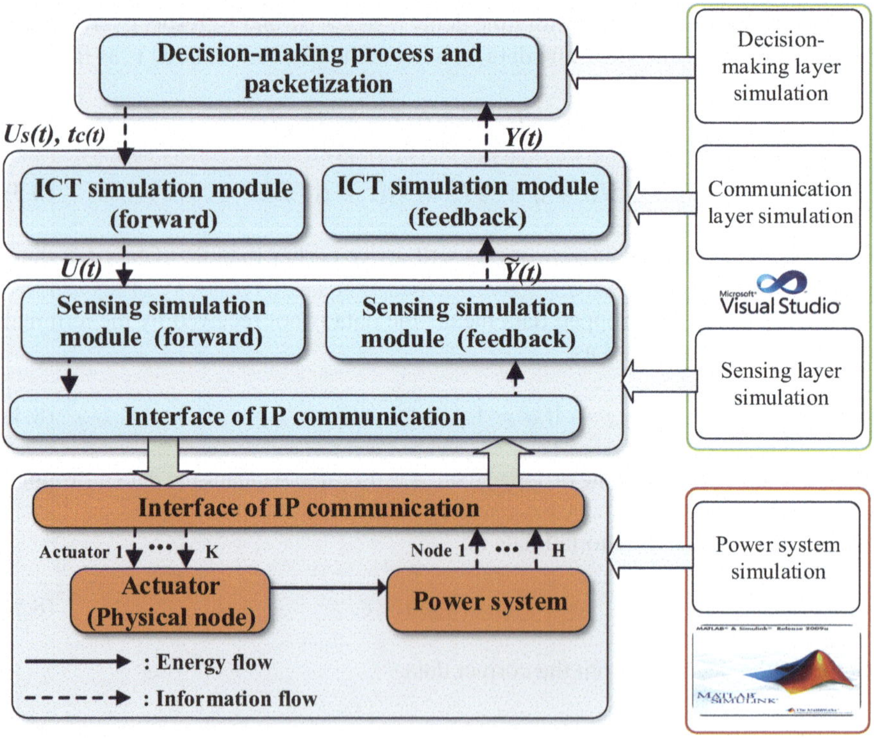

Fig. 8.2 Co-simulation block diagram

an important factor, and the time is also calculated in the decision-making layer simulation.

The IP-based Ethernet communication function is set up in both MATLAB and Microsoft Visual Studio to transfer data between different software programs. There is another reason to employ the IP-based Ethernet connection that the co-simulation model can be utilized to perform the cyber-contingency assessment in the practical communication network.

Sensing and Communication Layers' Simulation: The design of sensing and communication layers is configured in Microsoft Visual Studio. As mentioned in Sect. 8.2.2, it can be concluded that affected information flows possess two kinds of properties: data (distorted data) and time (delayed, disordered and dropped data). In the chapter, the transmitted information flow is modeled and simulated by realizing these two properties. Moreover, the impact of data and time properties is realized in the sensing layer simulation and communication layer simulation, respectively.

(1) *Sensing Layer Simulation:* The sensing layer simulation consists of the IP-based Ethernet communication module and the sensing simulation module. The former takes the responsibility of transferring packets between MATLAB and Microsoft Visual Studio, based on IP address, port number, and UDP protocol. The latter simulates the information flow of measurement devices and receivers of actuators. As shown in Fig. 8.2, the sensing simulation module includes feedback and forward loops. The functions of them are (1) sampling operation data from (feedback) or delivering decisions to (forward) corresponding actuators, (2) packetizing data (feedback), (3) implementing distorted data, which are shown in Fig. 8.3 The implemented distorted data is to cover distorted data in sensing, communication, and decision-making layers. The packetized data includes the device's ID, operation data/decision, and timestamp. The device's ID is used to identify the packets of sensing devices and receivers of actuators. The power system operation data (power, voltage, and so on) or decision made by the decision-making layer is stored in the operation data/decision. The timestamp records the sample time. The distorted data is generated by modifying values of the ID, operation data/decision, and timestamp, according to (8.5).

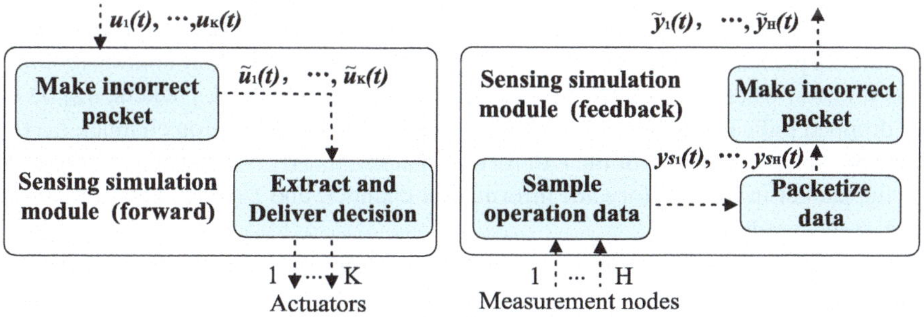

Fig. 8.3 Functions of sensing simulation module

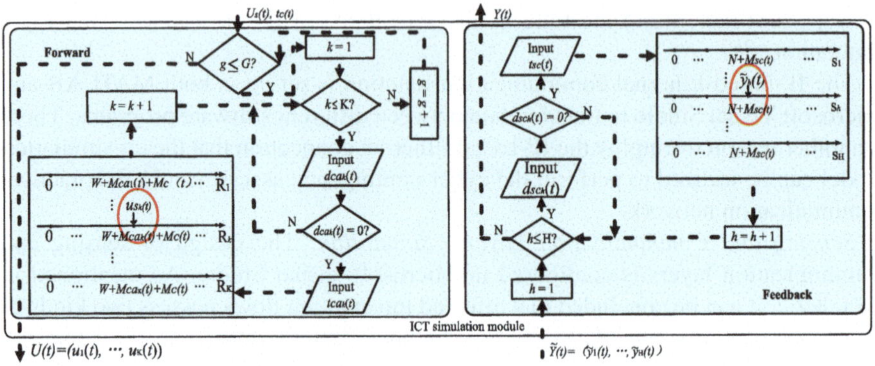

Fig. 8.4 Working flow of ICT simulation module

(2) *Communication Layer Simulation*: The ICT simulation modules in the communication layer simulation, as shown in Fig. 8.2, are to model end-to-end information flows between measurement devices and actuators and the decision-making center. The principle of ICT simulation modules is shown in Fig. 8.4.

The time delay utilizes the definition in [17], i.e., round-trip time (RTT) delay, shown as follows:

$$t_r = t_{sc} + t_c + t_{ca} \tag{8.6}$$

where t_r is the RTT delay, t_{sc} is the delay in the feedback process, t_{ca} is the delay in the forward process, and t_c is the computation time of the decision making.

Applications in the decision-making layer need to receive all feedback packets at the simulation time t to implement a decision, which means the packet having the worst time property will decide the property of packets at t. For example, the maximum delayed time of packets at t can be regarded as the delayed time of all packets at t. Therefore, feedback packets at t can be set as having the same time property. But for forwarding decision packets, different time properties will cause different behaviors of actuators. For example, a decision packet k with less delayed time than another packet $k + 1$ will reach the target actuator earlier, which causes the corresponding actuator to operate earlier. Hence, the different time properties of forwarding packets should be considered. The principle of the ICT simulation module is as follows, where $t_{sc}(t)$ is the pre-set delayed time of feedback packets, $d_{sch}(t)$ is the dropped indicator for the packet in the h th feedback information channel, $t_{cak}(t)$ is the pre-set delayed time of the k th forward packet, $d_{cak}(t)$ is the dropped indicator for the packet in the hth forward information channel, and t means the simulation time, respectively.

The following is procedures to simulate the time property of the transmitted information:

(1) Define the co-simulation time step ΔT and the sample time of power system applications. Then, initiate the memory region with default values for storing packets.

(2) The simulation time t is equal to the sample time, satisfying $t = N * \Delta T$ and N is a nonnegative integer.

(a) Packets of measurement devices from the sensing simulation module, whose number is H, i.e., $\tilde{Y}(t) = (\tilde{y}_1(t), \ldots, \tilde{y}_H(t))$, are sent to the ICT simulation module (feedback).

(b) For a packet, $\tilde{y}_h(t)$ in $\tilde{Y}(t)$, $h \in [1\ H]$, it is first to judge the value of the inputted $d_{\text{sch}}(t)$. If $d_{\text{sch}}(t) = 0$, $\tilde{y}_h(t)$ is stored at the $\{N + M_{\text{sc}}(t)\}$th position in the memory region, S_h, according to the inputted value of $t_{\text{sc}}(t) = M_{\text{sc}}(t) * \Delta T$, where $M_{\text{sc}}(t)$ is nonnegative integer. If $d_{\text{sch}}(t) = 1$, the packet is set as the dropped packet and the process of storing is ignored.

(c) Store $\tilde{Y}(t)$ at corresponding positions in the memory region, based on inputted values of $t_{\text{sc}}(t)$ and $D_{\text{sc}}(t) = (d_{\text{sc1}}(t), \ldots, d_{\text{scH}}(t))$.

(3) The arbitrary simulation time t satisfies $t = W * \Delta T$ and W is a nonnegative integer.

(a) The ICT simulation module (feedback) outputs $Y(t) = (S_1(W), \ldots, S_H(W))$ consisting of all stored packets together with their position information to the decision-making layer simulation. Because packets may be dropped and the difference between the sample time and simulation time, it is possible that some or all elements in the $Y(t)$ are the default value, which can be identified as the dropped data and default data by the decision-making layer simulation, respectively.

(b) Based on $Y(t)$, the decision-making layer simulation computes the decision set $U_s(t)$ and sends it to the ICT simulation module (forward) with the computation time $t_c(t)$. If $Y(t)$ consists of the default value, the decision computation is skipped.

(c) The computed decision set is $U_s(t) = (u_{s1}(t), \ldots, u_{\text{sK}}(t))$, where K is the number of actuators. For a packet, $u_{\text{sk}}(t)$, $k \in [1\ K]$, it is first to judge the value of the inputted $d_{\text{cak}}(t)$. If $d_{\text{cak}}(t) = 0$, $u_{\text{sk}}(t)$ is stored at the $\{W + M_{\text{cak}}(t) + M_c(t)\}$ th position in the memory region R_k, according to the inputted value of $t_{\text{cak}}(t) = M_{\text{cak}}(t) * \Delta T$ and corresponding computation time $t_c(t) = M_c(t) * T$, where $M_{\text{cak}}(t)$ and $M_c(t)$ are nonnegative integer. If $d_{\text{cak}}(t) = 1$, the packet is set as the dropped packet and the process of storing is ignored.

(d) Store $U_s(t)$ at corresponding positions in the memory region, based on inputted values of $T_{\text{ca}}(t) = (t_{\text{ca1}}(t), \ldots, t_{\text{caK}}(t))$, $D_{\text{ca}}(t) = (d_{\text{ca1}}(t), \ldots, d_{\text{caK}}(t))$ and the computation time $t_c(t)$.

(e) The ICT simulation module (forward) outputs $U(t) = (u_1(t), \ldots, u_K(t)) = (R_1(W), \ldots, R_K(W))$. And those packets act as practically prepared decision packets for controllable devices whose number is K. Because packets may be dropped and have different time delays in the forward loop, it is possible that some or all elements in the $U(t)$ are the default value, which can be identified and treated as no updated control decision by the sensing layer simulation, and the sensing layer will still send the last updated decision data to the power system simulation.

8.3.2 Time Synchronization and Data Exchange of Simulation

Time Synchronization: The power system simulation in MATLAB is time-driven, while the cyber system simulation programmed by C++ is event-driven. Hence, a time synchronization method is indispensable. Here, the master–slave synchronization method is adopted [18], as shown in Fig. 8.5. Power system simulation acts as the master and begins to work in advance of the cyber system simulation. At the defined co-simulation time step, ΔT, power system simulation halts and starts to drive the operation of cyber system simulations.

Data exchange of simulation: The working flow of the co-simulation is shown in Fig. 8.2. The data exchange procedure of co-simulation models at each synchronization time is as follows:

(1) At the sample time $t = N * \Delta T$, the power system simulation sends the operation data (i.e., rotor speed of a generator) to the cyber system simulation.
(2) Sensing layer simulation. Packetize the operation data with the physical equipment ID and sample time t as $Y_s(t)$. Then, make $Y_s(t) = (y_{s1}(t), \ldots, y_{sH}(t))$ become the distorted packet, $\tilde{Y}(t)$, by adding the pre-defined error $\varepsilon_{sc}(t) = (\varepsilon_{sc1}(t), \ldots, \varepsilon_{scH}(t))$ up on the operation data or timestamp. Then, transfer the packet, $\tilde{Y}(t)$, to the ICT simulation module (feedback).

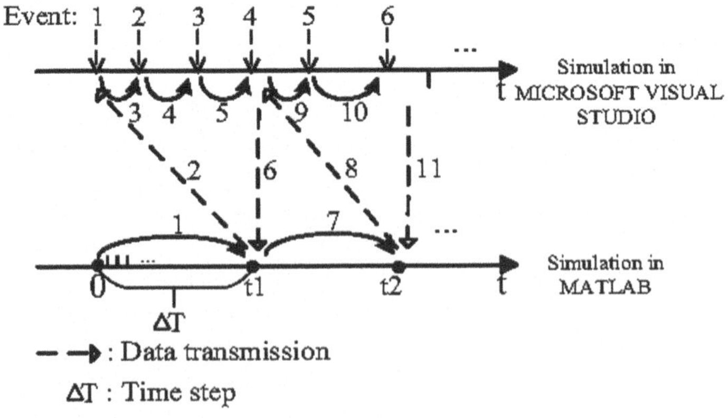

Fig. 8.5 Time synchronization mechanism

(3) Communication layer simulation. The ICT simulation module (feedback) achieves functions of generating delayed, dropped, and disordered packets, according to the pre-set $t_{sc}(t)$, $D_{sc}(t)$ and sends $Y(t)$ to the decision-making layer simulation at each time step.
(4) Decision-making layer simulation. The decision is made based on $Y(t)$, and the computation time, $t_c(t)$, of the decision-making process is calculated. Then, packetize the decision with the targeted physical controllable equipment ID and same timestamp of $Y(t)$ as the decision packet $U_s(t)$. At last, transfer the packet, $U_s(t)$, together with the computation time $t_c(t)$ to the communication layer simulation.
(5) Communication layer simulation. Receive packets, $U_s(t)$, and deliver them to the ICT simulation module (forward). Then, the ICT simulation module (forward) achieves functions of generating delayed, dropped, and disordered packets, according to the pre-set $T_{ca}(t)$, $D_{ca}(t)$, and the computation time $t_c(t)$. At last, send packets, $U(t)$, from the ICT simulation module to the sensing simulation module (forward).
(6) Sensing layer simulation. Make received packets become distorted packets $\tilde{U}(t)$, by adding the pre-defined error $\varepsilon_{ca}(t) = (\varepsilon_{ca1}(t), \ldots, \varepsilon_{caH}(t))$ up to the decision, ID, or the timestamp. Then, extract and deliver the decision from $\tilde{U}(t)$ to the controllable actuator corresponding to the ID in the packet.
(7) Power system simulation receives decisions and conducts controllable actuators.

8.3.3 Assessment of Cyber Contingencies

Compared with the physical contingencies assessment method, which analyzes the corresponding system response by scanning all possible physical contingencies, the assessment of cyber-contingencies (ACC) investigates physical system responses with and without cyber-contingencies. However, during the simulation of cyber contingencies, we only focus on simulating their impacts on the information flow.

The information features can be abstracted as delayed, disordered, dropped, and distorted data. Here, t_{sc} and T_{ca} are adopted to describe time delays of the data transmission in feedback and forward processes, respectively; D_{sc}, D_{ca} represent whether the data is dropped in feedback and forward processes, respectively; ε_{sc}, ε_{cs} represent the deviation of the distorted data from the correct data in feedback and forward processes, respectively; and the computation time of cyber applications in the cyber center is t_c.

The procedure of the assessment method is as follows, which can be seen in Fig. 8.1:

(1) Define the test system, which includes power and cyber systems.
(2) Generate the cyber-contingency set, which involves variations of t_{sc}, T_{ca}, D_{sc}, D_{ca}, ε_{sc}, and ε_{cs}. t_c is generated automatically in the cyber center simulation based on the application.

(3) Run the co-simulation, and at each simulation step, import values in the cyber-contingency set to the modules of sensing and ICT simulations.
(4) Obtain results of the co-simulation, and assess impacts of cyber-contingencies on physical system responses.

8.4 Case Studies

A modified IEEE benchmark system [19] with one shunt-type FACTS (SVC) device at the middle of the interconnected line, as shown in Fig. 8.6, is adopted in power system simulations. The system has a typical inter-area mode with $f = 0.674$ Hz and $\rho = 0.00619$.

WADC is employed to enhance the stability of interconnected power systems [19–22]. The primary goal of WADC is to dampen the low-frequency oscillation between areas in power systems. In this chapter, a model predictive control (MPC) [23] is utilized to design the wide-area damping control method developed in the decision-making layer simulation. The wide-area signal of the deviation of $P_{\text{line\#1}}$ is measured. The control output of MPC is transmitted and added up to the SVC as a supplementary

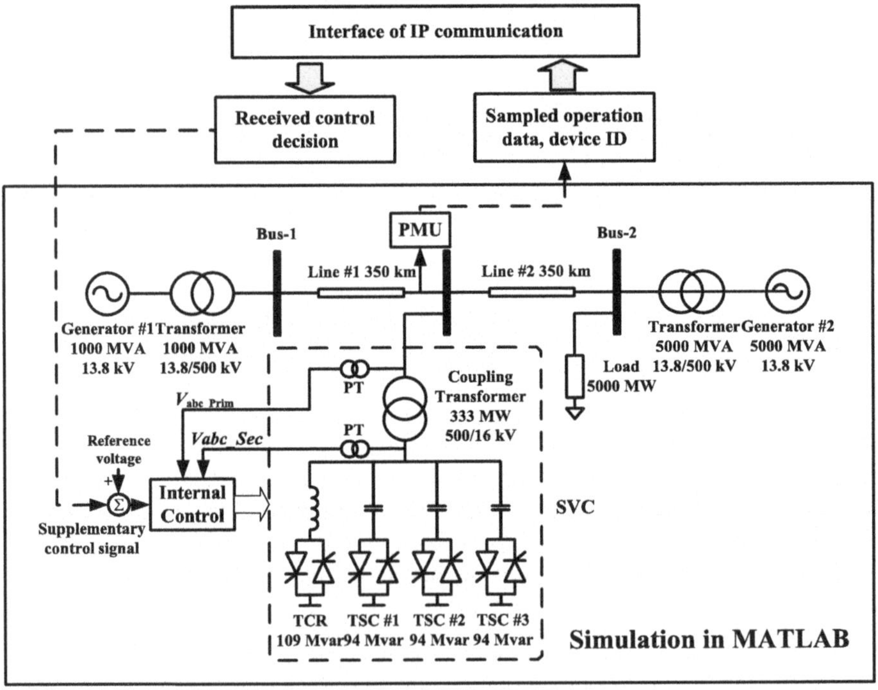

Fig. 8.6 Adopted modified IEEE benchmark system

control signal, which is limited within [−0.2, 0.2] pu. In MPC, the control interval, prediction horizon, control horizon, and reference are set as $\Delta t = 100$ ms, 25, 15, and 0, respectively. To assess the impacts of cyber contingencies on the operation of WADC, a three-phase-to-ground fault happening at 0.2 s is introduced to ignite the oscillation in power systems.

8.4.1 Verifying Simulation Method of Transmitted Data

Figure 8.7 shows co-simulation procedures of the conventional and proposed methods. In existing co-simulation platforms, the co-simulation procedures consist of defining parameters of cyber contingencies, getting impacted information flows by full process simulations of cyber systems, interacting with energy flows, and obtaining impacted energy flows. However, in the proposed method, the co-simulation procedures consist of defining parameters of information features, getting impacted information flows by the proposed data flow simulation principle, interacting with energy flows, and obtaining impacted energy flows. Through the above comparison, it can be found that there are two keypoints to decide accuracies of the proposed method: whether the specific cyber contingency can be represented by information features and whether the proposed simulation principle of impacted information flows is correct.

To check these two keypoints, three kinds of cyber contingencies, namely DoS, CLO, and MITM, are firstly simulated by the dedicated network simulator, OPNET, to get impacted transmission packets of the wide-area signal. Subsequently, via modeling information features identical with the packets simulated by OPNET, the impacted transmission packets are simulated by the proposed sensing layer and communication layer simulations. Finally, through comparing simulated impacted transmission packets of two methods, the accuracy of the proposed co-simulation method

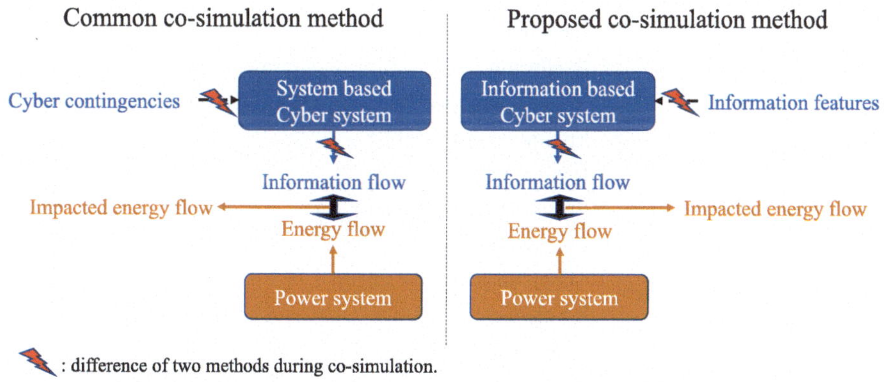

Fig. 8.7 Co-simulation procedures of conventional and proposed methods

can be verified. As verifications of the feedback and forward communications are similar, only the verification of feedback communication is implemented in the chapter, and the measurement value of transmission packets is adopted to show.

To begin with, a network model in OPNET, shown in Fig. 8.8, is set up for implementing communication functions. The main parameters of the network are shown in Table 8.1. Although the established cyber system is only a small-scale cyber

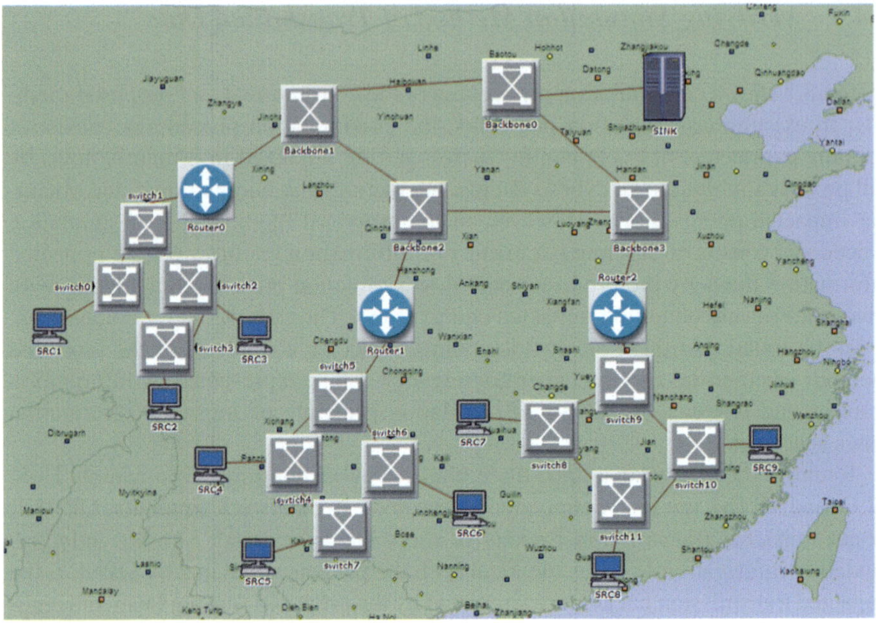

Fig. 8.8 Adopted cyber system architecture

Table 8.1 General cyber system settings

Variables	Parameters
Bandwidth	100 Mbps
Measurement rate	10 times/s
Transmission protocol	UDP
Routing protocol	RIP
Switch spanning tree protocol	RSTP
Interface protocol	CSMA/CD
Communication line length	About 2000 km
Router memory	512 MB
Router interface speed	10 Mbps
Router packet switching rate	500,000 packets/s
Router packet forwarding rate	100,000 packets/s

system, the complexity of modeling has already been shown, due to the architecture establishment and parameters setup. When the practical cyber system is large and supplies services to several power system applications, the number of cyber system parameters will surge, and some of them will be hard to be decided. Furthermore, any uncertainty on cyber system parameters will decrease the accuracy of the impact assessment. For example, when different bandwidths, protocols, and performances of routers are adopted in the cyber system, the effects of the DoS will be different. Additionally, for the CLO, the interruption time of the data will not be the same, when different switch spanning tree protocols are adopted.

Here, the wide-area signal of $\Delta P_{\text{line\#1}}$ is assigned to be transmitted to the cyber center at the node SINK, from the measurement node SRC 1. For demonstration, the sampling time and simulation time step are set as 100 ms and 50 ms, respectively. The received time of packets in OPNET is edited to the same simulation time step of 50 ms, by changing it to the time which is the smallest integer multiple of 50 ms, as well as, larger than or equal to the received time.

(1) DoS Attack: The DoS attack aims at exhausting communication resources of the target by generating useless communication traffics. Consequently, the target computer is not able to address users' legitimate requests. There are 8 potential nodes to launch the attack, namely SRC 2– 9. Besides, there are kinds of way to implement the DoS attack, such as UDP flood, SYN flood, land attack. Hence, the cyber-contingency set in conventional co-simulation methods needs to cover attacks from all potential nodes together with possible attack patterns.

In OPNET, the attacker at the compromised measurement node SRC 5 adopts the UDP flood and floods the communication interface on the control center server, SINK, by sending the useless data of 60 Mbps. Due to the DoS attack, traffic congestions happen on the control center server, and the communication delay of the wide-area signal continuously increases. Figure 8.9 shows received packets of $\Delta P_{\text{line\#1}}$ at the cyber center. The solid is sampled operation status of power systems without cyber attacks. The stems represent received packets simulated by OPNET. The stairs are

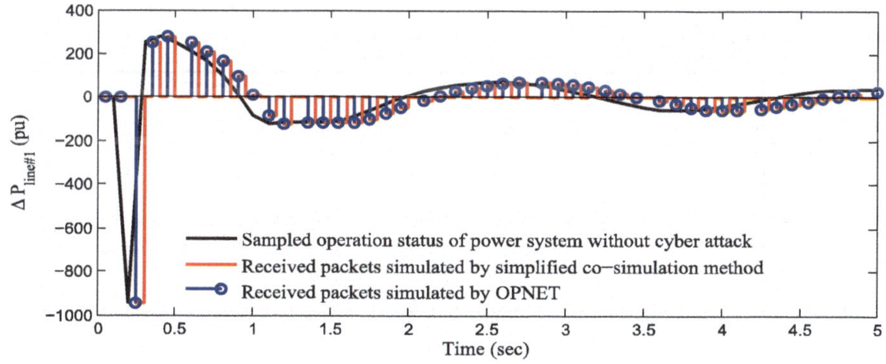

Fig. 8.9 Received packets of $\Delta P_{\text{line\#1}}$ under DoS attack

the received packets simulated by the proposed method. In the proposed method, the information feature of the delayed data is defined based on the communication time delay of packets simulated by OPNET. In the results of the proposed model, the rising edge means the new packet is received. And then, the new packet holds for a time step ΔT. It is evident that the packets of OPNET and the proposed method are refreshed to the same value simultaneously. It can be then verified that through modeling the delayed data, received packets simulated by the proposed method coincide with ones of OPNET simulation. Thus, the DoS attack can be represented by the information feature of the delayed data, and the simulation principle of the delayed information flow is accurate.

Besides, unlike the cyber-contingency set in conventional co-simulation methods, the information feature-based cyber-contingency set just needs to define the delayed data in the end-to-end information flow between the SRC 1 and SINK. Moreover, besides this attack pattern of the DoS, cyber attacks which generate similar delayed impacts on the end-to-end information flow can be merged.

(2) CLO Attack: The CLO attack results in variations of the communication network structure. The switch spanning tree protocol needs a period to converge the communication network. Before the network is converged, the packets going through the interrupted communication line will be dropped. There are 12 communication links related to the communication from SRC 1 to SINK, and the one-link interruption happens in 8 of them will not result in the permanent data loss. Hence, the cyber-contingency set in conventional co-simulation methods needs to cover attacks at all potential links.

OPNET simulates the CLO, happening at 2 s, between nodes backbone 0 and backbone 1. Because the interrupted line is utilized by communications between the node SRC 1 and node SINK, the packets between these two nodes will be dropped, after the outage happens. Figure 8.10 shows received packets of $\Delta P_{\text{line\#1}}$ at the cyber center. The stems are received packets simulated by OPNET. There is a time

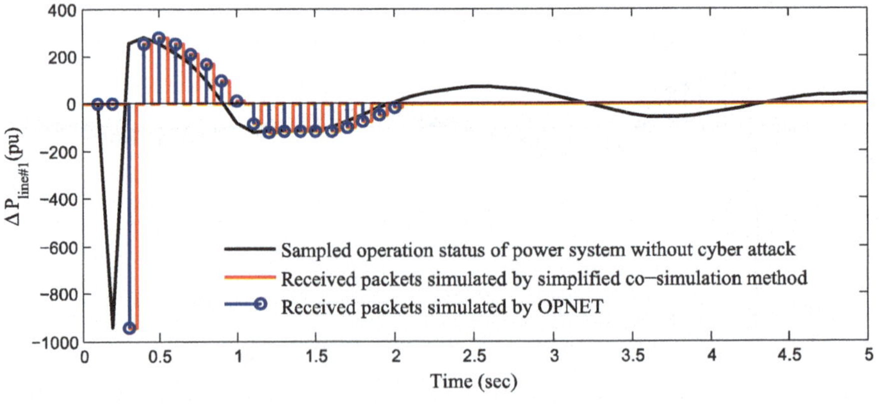

Fig. 8.10 Received packets of $\Delta P_{\text{line\#1}}$ under CLO

delay of 100 ms ($2 * \Delta T$) due to the remote communication. The stairs represent received packets simulated by the proposed method. In the proposed method, the dropped data is defined based on the packets simulated by OPNET. It can be seen that, before the outage happens, the communication is healthy and $\Delta P_{\text{line\#1}}$ of two methods are coincident. When the CLO occurs, received packets of OPNET and the proposed method are interrupted at the same time, which shows the CLO attack can be represented by the information feature of the dropped data, and the simulation principle of the dropped information flow is accurate. Besides, unlike the cyber-contingency set in conventional co-simulation methods, the information feature-based cyber-contingency set just needs to set the dropped data in the end-to-end information flow between the SRC 1 and SINK. Moreover, besides the CLO, cyber attacks which generate similar dropped impacts on the end-to-end information flow can be merged.

(3) MITM Attack: The MITM attack is utilized to build up an unknown connection with both endpoints of targeted communication and maintain the communication between them, so victims still believe that their communication is safe and regular. In CPEPS, a successful MITM attack between the cyber center and measurement devices and actuators will give the attacker the same vision of system operators. Meanwhile, measurement values, control command, and price policy, etc., transmitted in cyber systems can also be falsified by the attacker. There are 8 potential nodes to launch the attack, namely SRC 2–9. Besides, the attack patterns of MITM include ARP spoofing, DNS spoofing, session hijack, and so on. Hence, the cyber-contingency set in conventional co-simulation methods needs to cover attacks from all potential nodes together with possible attack ways.

In OPNET, the attacker manipulates the routing table of the Router 0 to make the vicious computer at SRC 3 stay in the connection between the measurement node SRC 1 and center node SINK. Thus, the attacker can stealthily stay between the local measurement node and a cyber center. We assume the attacker manipulates transmission packets by increasing 30% of the measurement value. Received packets of $\Delta P_{\text{line\#1}}$ at the cyber center are shown in Fig. 8.11 The stems are received packets simulated by OPNET. Like the simulation in the CLO, there is also a time delay of 100 ms ($2 * \Delta T$) induced by the remote communication. The stairs are received packets simulated by the proposed method. In the proposed method, the distorted data is defined based on packets simulated by OPNET. It can be found that received packets simulated by the proposed information flow simulation method are identical with packets simulated by OPNET, which shows the MITM attack can be represented by the distorted data, and the simulation principle of the dropped information flow is accurate. Besides, unlike the cyber-contingency set in conventional co-simulation methods, the information feature-based cyber-contingency set just needs to set the distorted data in the end-to-end information flow between SRC 1 and SINK. Moreover, besides the MITM, cyber-attacks which generate similar distorted impacts on the end-to-end information flow can be merged.

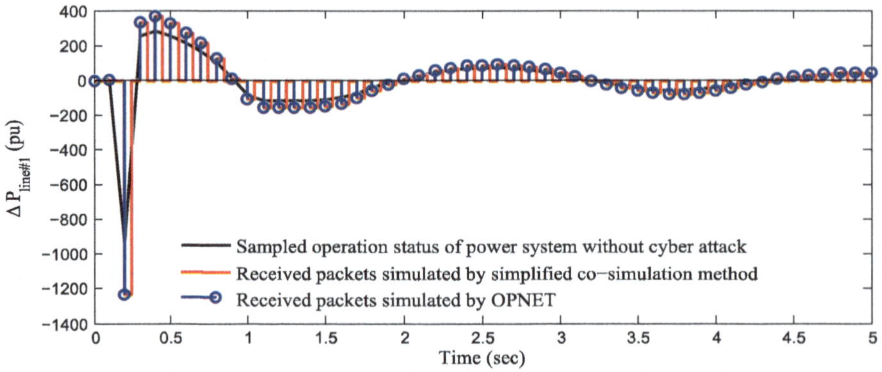

Fig. 8.11 Received packets of $\Delta P_{\text{line\#1}}$ under MITM attack

8.4.2 Cyber-Contingency Assessment

Cyber contingencies will cause the delayed, disordered, dropped, and distorted data, respectively, or result in the combination of some of them. On the one hand, the definition of these information features can be performed based on historical records. On the other hand, via defining information features, operations of CPEPS under some desire scenarios can be analyzed. In this chapter, the impacts of the delayed, disordered, dropped, and distorted data will be analyzed, respectively. Their combined impacts can be investigated by defining combinations of the following cyber-contingency sets.

For the demonstration, the cyber-contingency sets are as follows:

(1) Delayed and Disordered Data: random time delay in both feedback and forward processes with the same range of 100 ms to 300 ms, 200 ms to 400 ms, and 300 ms to 500 ms, respectively.
(2) Dropped Data: a data interruption of one second in the forward channel is adopted. Three time zones are employed, namely 0.2–1.2 s, 1–2 s, and 2–3 s, which can represent the data interruption happening at the early, middle, and last stage of the damping process, respectively.
(3) Distorted Data: add up 5 pu and 10 pu to the wide-area signal, respectively.

Impacts of Delayed Data: Figure 8.12 shows the impacts of the delayed data caused by different ranges of the random time delay, and $d\omega_{12}$ in the figure means the rotor speed deviation of generators #1 and #2. The rising time delay results in the oscillation happening in physical power systems, and Fig. 8.13 shows timestamps of implemented control decisions. The timestamp indicates the sample time of wide-area signals on which the control decision is made. With the increasing time delay, the difference between the timestamp of control decisions and simulation time is growing.

It means there is an increasing gap between the physical operation status on which the control decision is made and the actual operation status when the control decision

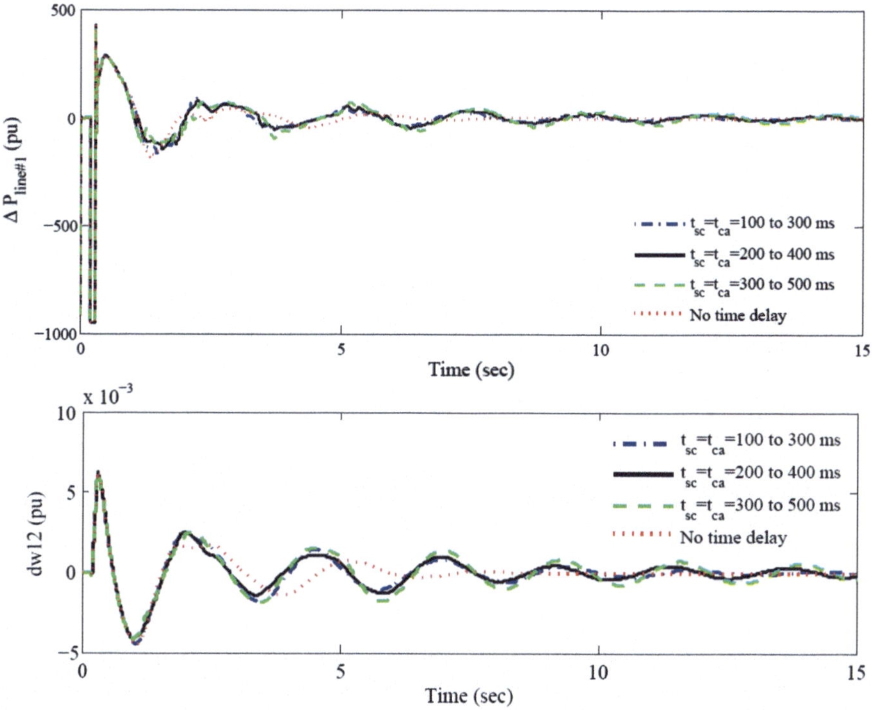

Fig. 8.12 Damping performance under different ranges of random time delay

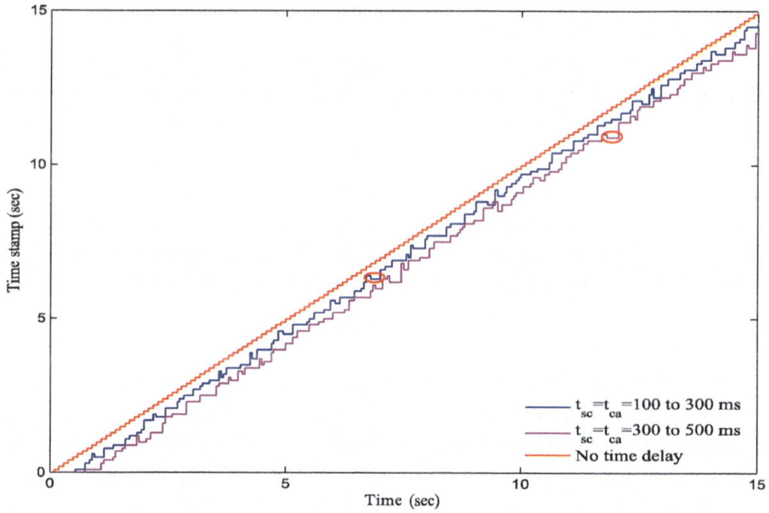

Fig. 8.13 Timestamp of implemented control decisions at each control interval under random time delay

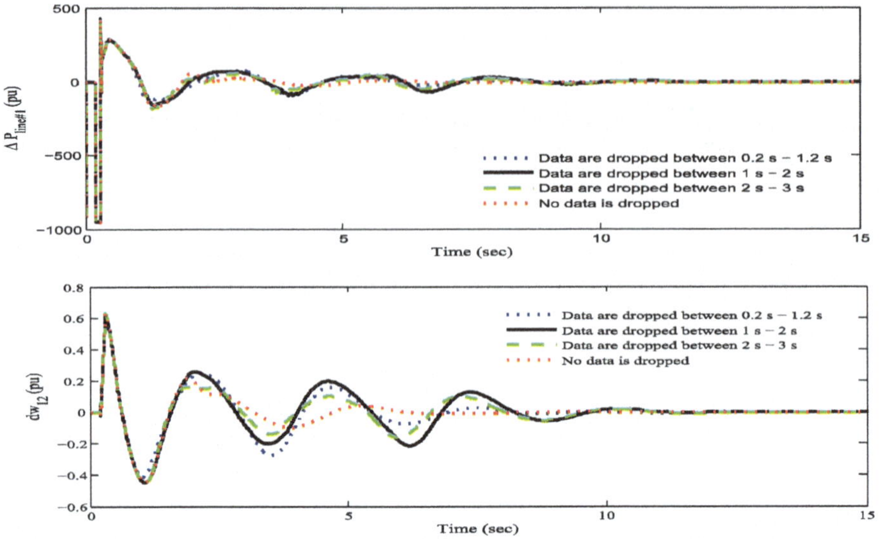

Fig. 8.14 Damping performance under different packet drop rates

is implemented, which will significantly deteriorate physical damping performances. Besides, the disordered data caused by the random time delay is partly marked by the red circle in Fig. 8.13, where the timestamp decreases compared to the last timestamp. The invariable timestamp of adjacent control intervals is because the SVC implements the control decision received before, when the control decision at the current simulation time is absent.

Impacts of Dropped Data: Figure 8.14 shows the impacts of the dropped data on physical damping performances. When the data is dropped, the cyber center cannot evaluate the power system status and generate control decisions, and therefore, it will be also absent that the control commands received by actuators in the physical process. From the figures, it can be found that the data interruption happening in the middle stage of the damping process generates the worst impact on the information flow. Figure 8.15 shows implemented control decisions. It can be discovered that the dropped data results in the failed control decision updating in the SVC; thus, the SVC implements the control decision received before, which will inevitably deteriorate physical damping performances.

Impacts of Distorted Data: State estimation is an essential application in power systems for obtaining system operation conditions and acts as a necessary counteraction for identifying the manipulated data [24–26]. However, adversaries can utilize the well-designed attack strategy to implement an unobservable assault, which means the manipulated data cannot be detected by either DC or AC state estimation [27, 28]. Besides, the state estimation in the power system is generally performed in the scale of minutes. However, the low-frequency oscillation damping is in the scale of seconds. The state estimation is hard to identify the distorted data in time. Here, we

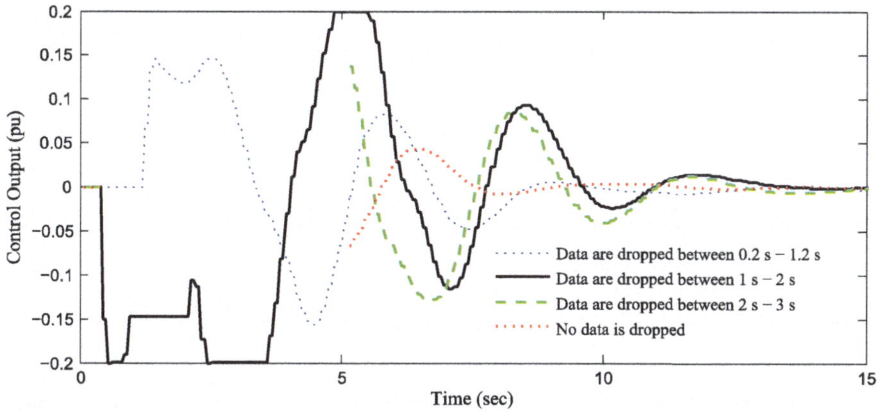

Fig. 8.15 Implemented control decision at each control interval under dropped data

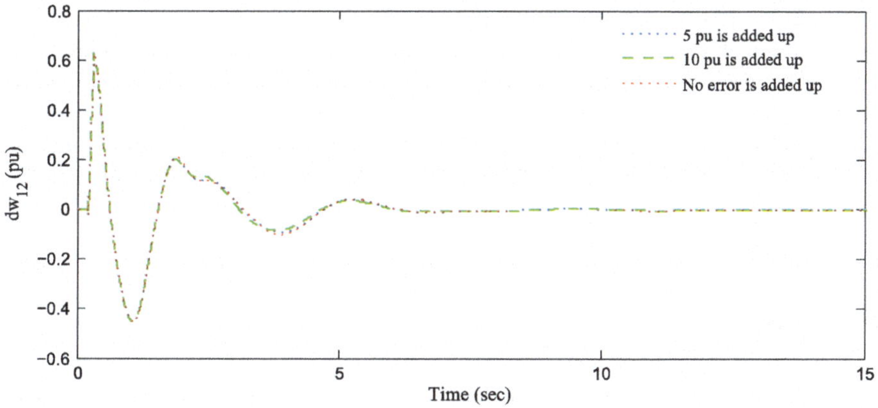

Fig. 8.16 Physical damping performances under different manipulated operation data

assume that attackers perform a successful unobservable attack, and the manipulated wide-area signal $\Delta P_{\text{line\#1}}$ cannot be detected by the state estimation. From Figs. 8.16 and Fig. 8.17, it can be seen that, although the constant distorted data of $\Delta P_{\text{line\#1}}$ only generates slight impacts on damping performances, the control decisions for the FACTS are changed, which alters the output of the SVC.

8.5 Summary

This chapter discusses categories of cyber contingencies and their impacts on information flows. Through modeling end-to-end information features and simulating impacted information flows, an information-based co-simulation model is given,

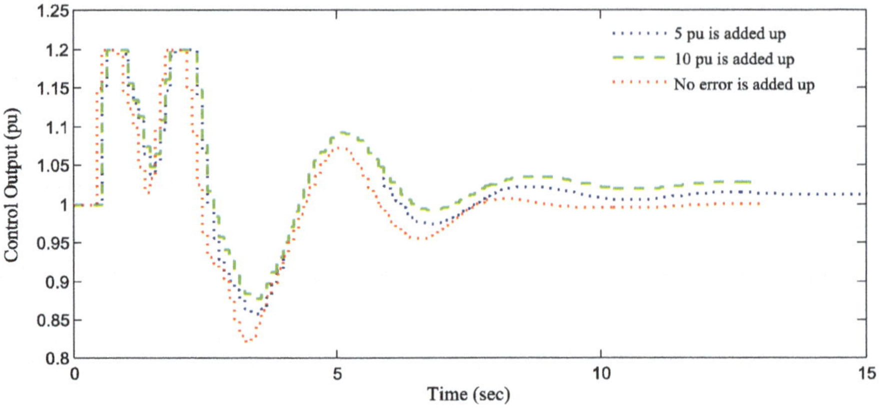

Fig. 8.17 Implemented control decisions of FACTS under different manipulated operation data

which can be used to investigate the impacts of cyber contingencies on power systems. The developed co-simulation model can reduce the modeling complexity of cyber systems and avoid impacts of cyber systems uncertainties on the cyber contingencies' assessment. Moreover, the scale of considered contingencies can be reduced by merging various contingencies based on their impacts on the information flow. Besides, the cyber contingencies' assessment can quantify the relation between the cyber and physical systems and provide the guideline for operations of CPEPS.

References

1. Ilic M, Xie L, Khan U, Moura J (2010) Modeling of future cyber-physical energy systems for distributed sensing and control. IEEE Trans Syst Man Cybern A Syst Humans 40(4):825–838
2. Yao W, Jiang L, Wen J, Wu Q, Cheng S (2015) Wide-area damping controller for power system interarea oscillations: a networked predictive control approach. IEEE Trans Control Syst Technol 23(1):27–36
3. Natori K, Tsuji T, Ohnishi K, Hace A, Jezernik K (2010) Time-delay compensation by communication disturbance observer for bilateral teleoperation under time-varying delay. IEEE Trans Ind Electron 57(3):1050–1062
4. Chaojun G, Jirutitijaroen P, Motani M (2015) Detecting false data injection attacks in AC state estimation. IEEE Trans Smart Grid 6(5):2476–2483
5. Yu Z, Chin W (2015) Blind false data injection attack using PCA approximation method in smart grid. IEEE Trans Smart Grid 6(3):1219–1226
6. Befekadu G, Gupta V, Antsaklis P (2015) Risk-sensitive control under Markov modulated denial-of-service (DoS) attack strategies. IEEE Trans Autom Control 60(12):3299–3304
7. Zhang H, Cheng P, Shi L, Chen J (2015) Optimal denial-of-service attack scheduling with energy constraint. IEEE Trans Autom Control 60(11):3023–3028
8. Cao Y, Shi X, Li Y, Tan Y, Shahidehpour M, Shi S (2017) A simplified co-simulation model to investigate impacts of cyber-contingency on power system. IEEE Trans Smart Grid 9(5):4893–4905
9. Hahn A, Ashok A, Sridhar S, Govindarasu M (2013) Cyber-physical security testbeds: architecture, application, and evaluation for smart grid. IEEE Trans Smart Grid 4(2):847–855

10. Liu R, Vellaithurai C, Biswas S, Gamage T, Srivastava A (2015) Analyzing the cyber-physical impact of cyber events on the power grid. IEEE Trans Smart Grid 6(5):2444–2453
11. Sun X, Chen Y, Liu J, Huang S (2014) A co-simulation platform for smart grid considering interaction between information and power systems. In: Proceedings of the IEEE PES Innovations Smart Grid Technologies Conference (ISGT), Washington, DC, USA, February 2014, pp 1–6
12. Kazmi J, Latif A, Ahmad I, Palensky P, Gawlik W (2016) A flexible smart grid co-simulation environment for cyber-physical interdependence analysis. In: Proceedings of the Workshop on Modeling and Simulation Cyber-Physical Energy Systems (MSCPES), Vienna, Austria, April 2016, pp 1–6
13. Wei M, Wang W (2014) Greenbench: a benchmark for observing power grid vulnerability under data-centric threats. In: Proceedings of the IEEE Conference Computer Communication (INFOCOM), Toronto, ON, Canada, April 2014, pp 2625–2633
14. Venkataramanan V, Srivastava A, Hahn A (2016) Real-time co-simulation testbed for microgrid cyber-physical analysis. In: Proceedings of the Workshop Model Simulation Cyber-Physical Energy Systems (MSCPES), Vienna, Austria, April 2016, pp 1–6
15. Xin S, Guo Q, Sun H, Zhang B, Wang J, Chen C (2015) Cyber-physical modeling and cyber-contingency assessment of hierarchical control systems. IEEE Trans Smart Grid 6(5):2375–2385
16. Lin H, Deng Y, Shukla S, Thorp J, Mili L (2012) Cyber security impacts on all-PMU state estimator—A case study on cosimulation platform GECO. In: Proceedings of the IEEE 3rd International Conference on Smart Grid Communications (SmartGridComm), Tainan, Taiwan, November 2012, pp 587–592
17. Hu W, Liu G, Rees D (2008) Networked predictive control over the Internet using round-trip delay measurement. IEEE Trans Instrum Meas 57(10):2231–2241
18. Shi X, Li Y, Cao Y, Tan Y (2015) Cyber-physical electrical energy systems: challenges and issues. CSEE J Power Energy Syst 1(2):36–42
19. Li Y, Zhou Y, Liu F, Cao Y, Rehtanz C (2017) Design and implementation of delay-dependent wide-area damping control for stability enhancement of power systems. IEEE Trans Smart Grid 8(4):1831–1842
20. Li Y, Liu F, Cao Y (2015) Delay-dependent wide-area damping control for stability enhancement of HVDC/AC interconnected power system. Control Eng Pract 37:43–54
21. Yao W, Jiang L, Wen J, Wu Q, Cheng S (2014) Wide-area damping controller of FACTS devices for inter-area oscillations considering communication time delays. IEEE Trans Power Syst 29(1):318–329
22. Liu Y, Wu Q, Kang H, Zhou X (2016) Switching power system stabilizer and its coordination for enhancement of multi-machine power system stability. CSEE J Power Energy Syst 2(2):98–106
23. Azad S, Iravani R, Tate J (2013) Damping inter-area oscillations based on a model predictive control (MPC) HVDC supplementary controller. IEEE Trans Power Syst 28(3):3174–3183
24. Kim S, Wang G, Giannakis G (2014) Online semidefinite programming for power system state estimation. In: Proceedings of the IEEE Conference on Acoustics, Speech and Signal Processing, Florence, Italy, May 2014, pp 6024–6027
25. Wang G, Kim S, Giannakis G (2014) Moving-horizon dynamic power system state estimation using semidefinite relaxation In: Proceedings of the IEEE PES General Meeting, National Harbor, MD, USA, July 2014, pp 1–5
26. Zonouz S, Rogers K, Berthier R, Bobba R, Sanders W, Overbye T (2012) SCPSE: security-oriented cyber-physical state estimation for power grid critical infrastructures. IEEE Trans Smart Grid 3(4):1790–1799
27. Sun Y, Li W, Song W, Yuen C (2015) False data injection attacks with local topology information against linear state estimation. In: Proceedings of the IEEE Innovations Smart Grid Technologies Asia (ISGT ASIA), Bangkok, Thailand, November 2015, pp 1–5
28. Liang J, Kosut O, Sankar L (2014) Cyber attacks on AC state estimation: unobservability and physical consequences. In: Proceedings of the IEEE PES General Meeting Conference Expo, National Harbor, MD, USA, July 2014, pp 1–5

Chapter 9
JADE-Based Information Physical System Co-simulation Environment for Smart Distribution Networks

With the development of power electronics and communication technology, more distributed generators, energy storage devices, and electric vehicles are connected to the distribution network and bring many problems, such as multi-direction power flow and three-phase unbalance. To solve these problems, many methods based on advanced Information and Communications Technology (ICT) and multi-agent system (MAS) have been proposed [1, 2]. Compared with traditional methods, these new methods can be applied to large-scale complex distribution networks and have fast response speed and strong anti-interference ability. However, they also have the disadvantages of relying on communication and high cost. In order to verify its effectiveness and reduce the risk, the co-simulation environment became the preferred solution.

Co-simulation, combining power system and communication system, can clearly show the interaction between energy flow and information flow of smart distribution networks. It can not only analyze the physical state of power flow and transient fault of power grid, but also simulate the events of packet loss, delay, and network attack of communication system, so as to make the simulation results more comprehensive and reliable. Many co-simulation environments have emerged in recent years. Proposed by the US Air Force Technical College, EPOCHS is the earliest co-simulation environment [3, 4], and it supports the connection of PSCAD/EMTDC and NS2. In the latest version, AgentHQ interface is added to support multi-agent system. However, due to the synchronization delay caused by its fixed-step synchronization mechanism, the accuracy of large-scale system simulation would be affected; VPNET uses coordination software to synchronize and exchange data between VTB and OPNET [5]. Because the coordination software has no open interface, it is impossible to support multi-controllers function. ADEVS independently developed power simulation software independently and controlled the communication simulation software of NS2 [6], but its control system and power simulation were developed together without considering distributed control. On the contrary, PowerNet and GECO, based on communication simulation software NS2, control power simulation software Modelica and PLSF respectively to realize co-simulation [7, 8]. Power system control algorithms are implemented in power simulation software and cannot be designed separately.

Y. Cao et al., *Cyber-Physical Energy and Power Systems*,
https://doi.org/10.1007/978-981-15-0062-6_9

C2WT and NCSWT are developed from the combat simulation environment of the US military [9, 10]. Simulink is used to connect with NS2 and OMNeT++ to realize co-simulation, respectively. They are based on high-level architecture (HLA) and have good expansibility, but because of the military background of these two environments, the existing models support for power system simulation are not enough. INSPIRE developed by Technische Universität Dortmund is also based on HLA [11, 12]. It uses PowerFactory and OPNET to carry out co-simulation, and develops its own controller model, which can realize the simulation of distributed system control. However, because the controller model has not a standard interface definition, the implementation is more complex. In addition, there are some co-simulation environments, such as SCADASim [13], SmartSecLab [14], and TASSCS [15], which focus on the reliability and security of communication systems. The power simulation in these environments only analyzes the steady-state model, but they cannot fully reflect the changes of power system caused by distributed controllers. The co-simulation method based on JADE [16], which fits for the simulation of distributed system and multi-agent system, can simulate the physical and the communication models concurrently.

Most of the current co-simulation environments do not consider the impact of the controller on the power grid, and the simulation of the interaction between the controllers is weak. At the same time, it is not convenient to apply the library functions based on MAS in the co-simulation, which greatly reduces the effectiveness of the simulation and limits the simulation and application of MAS in the power system. These issues require a further study.

9.1 Distributed Control Joint Simulation Environment for Distribution Network

9.1.1 Architecture

The proposed architecture of co-simulation environment is shown in Fig. 9.1. The main body of the environment is embedded in a JAVA program based on JADE framework. In the program, there are many agents to perform a variety of tasks. PowerAgent is responsible to interact with the power system simulation software; NetAgent is in charge of the communication simulator. Agent1–AgentN are various types of controller in accordance with the design requirements. If an actual controller is connected to the co-simulation environment, the HILAgent is responsible for connection. All of these agents are managed by the agent Coordinator (i.e., RTI in HLA) to advance time, process data, and send messages.

Due to the powerful ability of simulation and analysis, this software has been widely used in modeling, simulation, and analysis of distribution network with renewable generations such as wind and solar power generation as well as the energy storage devices [17]. By means of the Hybrid Programming with JAVA and C++,

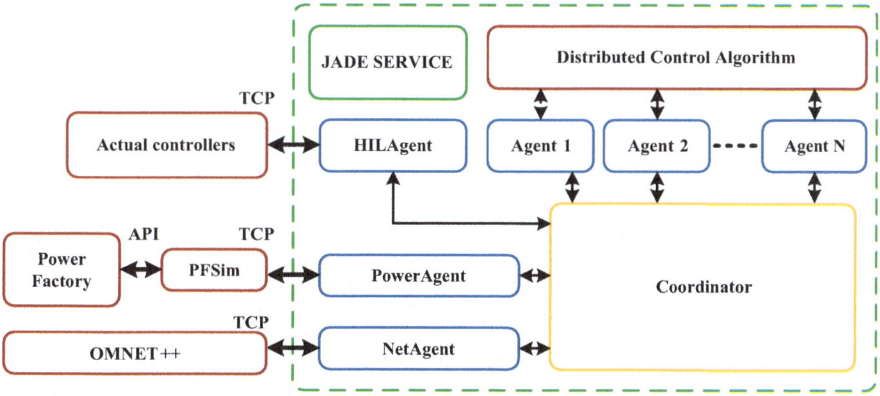

Fig. 9.1 Proposed architecture of co-simulation environment

PowerAgent can configure and control the PowerFactory software by calling API functions.

The choice of communication simulation is OMNeT++. As discrete event simulation software, OMNeT++ can simulate a network from different levels and details [18]. Compared with NS2, this software has a commercial version that you can upgrade and get better service if necessary. All of the simulation operation is implemented within the OMNeT++ based on C++ language. The NetAgent and the OMNet++ communicate via TCP/IP protocol.

9.1.2 Time Synchronization Mechanism

Simulation loop is executed to end in accordance with the following steps:

Step 1: Each member of the simulation calculates its own Lookahead Time. Lookahead Time indicates that the federation members will not produce new events in the future "Lookahead" time, so RTI can send and receive messages to the members of the federation concurrently within this time window.

Step 2: After that all of members sent their Lookahead Time to RTI, the RTI calculates the Lower Bound Time Stamp (LBTS). Since all simulation members are "Logical time synchronized," the equation for the LBTS is

$$\mathrm{LBTS} = \mathrm{Min}(T_i + \mathrm{Lookahead}_i) \quad i = 1, 2, \ldots, n \tag{9.1}$$

where T_i is the current logical time of federate i; $\mathrm{Lookahead}_i$ is the Lookahead Time of federate i.

Step 3: RTI advances the global time to the LBTS; then each member in federation runs simulation and handles its own events in message queue, respectively.

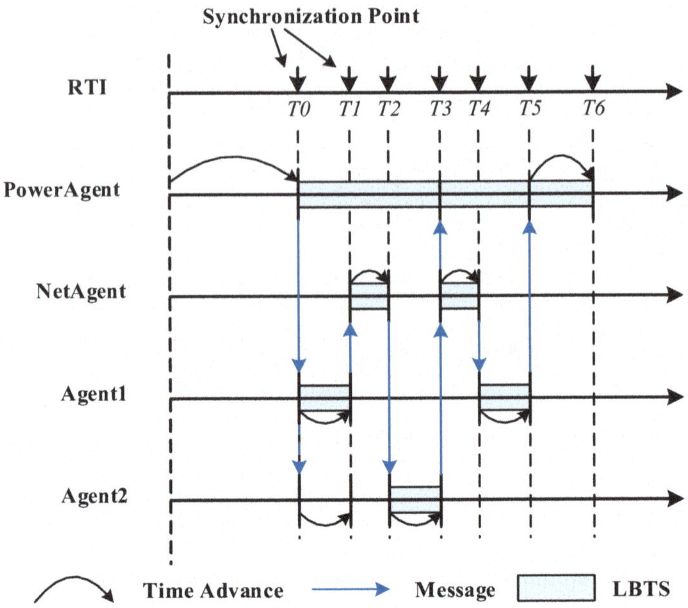

Fig. 9.2 Time synchronization of co-simulation environment

Step 4: RTI determines if the simulation reaches the end time. If not, then returns to *Step* 1.

Figure 9.2 describes the time synchronization in the distributed protection for smart distribution network. Assume the current logical time is T_0. After PowerAgent sends the grid simulation data to Agent1 and Agent2, these two controllers begin to process data and then all members calculate their Lookahead Time. The time of PowerAgent is T_6; Agent1 is T_1 since it has detected the overcurrent. Because Agent2 is close to the load side, there is no fault detected, so the Lookahead Time of Agent2 is infinity as well as the NetAgent. By comparison, RTI selects T_1 as LBTS and advances the time to T_1. At the logical time T_1, NetAgent simulates the communication from Agent1 to Agent 2, and the LBTS is T_2 at this time. Then RTI advances the time to T_2. The same work is repeated until the LBTS reaches the end time.

9.1.3 Processing of Event Chain

When Agents deal with their own events, they often need the assistance of other agents, so they send messages to related agents. At the same time, agents which get these messages would need help and will send messages to more agents. From this, we can see that agent's event processing is not dealing with a message, but dealing

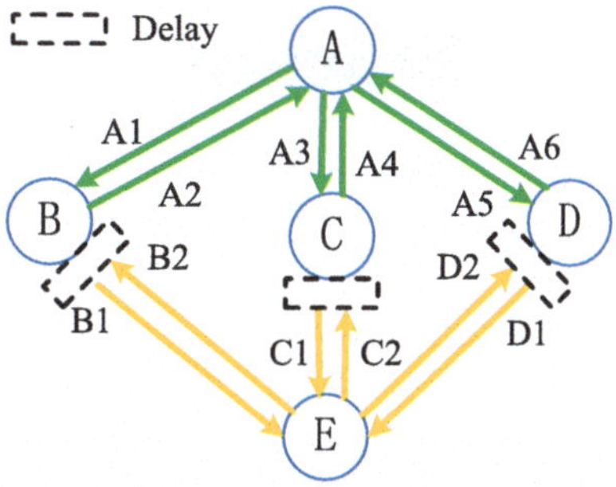

Fig. 9.3 Processing of event chain

with the state changes of all relevant agents brought by the message and the new messages generated by these changes. Because the above process is similar to tree traversal in data structure, it is easy to process event chain by recursive method. However, due to the circular transmission of messages, which will lead to the event chain processing into a dead-cycle state, additional mechanisms need to be added to prevent the generation of loops. The co-simulation environment uses the response delay of the controller and the transmission delay of the communication system to solve this problem.

As shown in Fig. 9.3, when PowerAgent (A) receives the activation message, it begins to read the voltage, current value, and switching status of each node in PowerFactory, and then sends the data to each controller agent (B, C, and D) in the order of A1 to A6. If the controller detects that the current exceeds the current threshold, it needs to send a fault message to the server (E), but because of the response time of the controller, the message will be stored until the response time elapses. On the one hand, the addition of response delay reflects the real characteristics of the controller, and on the other hand, it also interrupts the event chain and prevents the occurrence of message circular transmission.

9.2 Description of the Design Methods in Distributed Controllers

9.2.1 Simulation Environment of Distributed Controller

JADE framework comprises one or more containers. These containers can run on a computer, or on any equipment that support JAVA virtual machine, but there is only one main container at the same time. The messages in the framework are located and passed automatically via IMTP. All the processes are transparent to the application. This feature made JADE has excellent scalability and cross-platform capability, and it is easy to add new components as well as actual controllers to the environment.

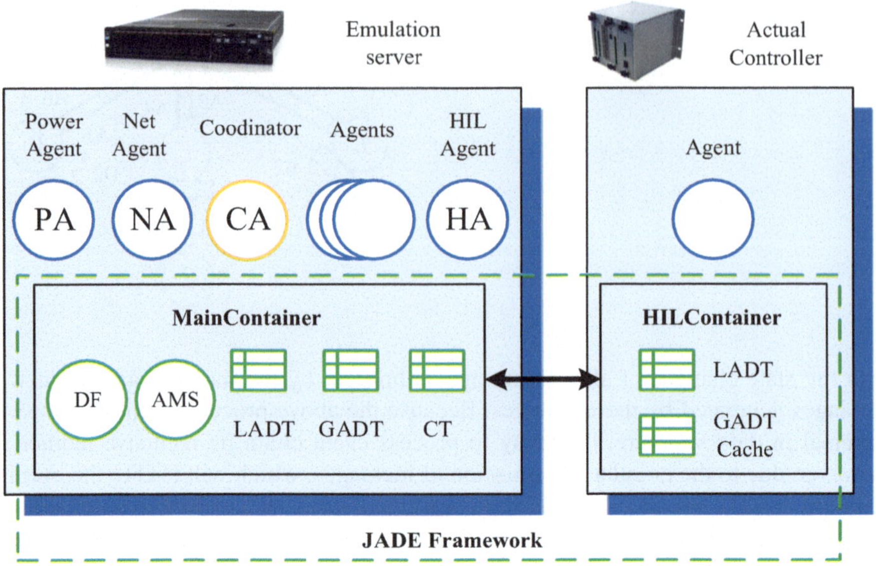

Fig. 9.4 Proposed simulation environment of distributed controller

As shown in Fig. 9.4, JADE container consists of two parts: system information tables and system services [19]. Container Table (CT) is used to record the information of all containers; global agent descriptor table (GADT) and local agent descriptor table (LADT) describe information of all agents in JADE environment and in local container, respectively. These three tables provide all information for JADE environment to operate the distributed system. Agent Management System (AMS) can provide monitoring and control services for tasks which want to handle agents, and directory facilitator (DF) adds the yellow pages service for the system. These two services provide good support for coordination and decision making of multi-agent system.

9.2.2 Implementation of Controller

In the smart distribution network, there are different controllers with different tasks. Such as the microcomputer relay protection equipment in substation and the feeder terminal unit on electric line pole, there are many differences between them in computing speed, sampling rate, and so on. When these devices run in the same system, their differences will make an enormous impact on distributed control. In this chapter, the proposed co-simulation environment reveals these features of controllers by the properties shown in Table 9.1.

Table 9.1 Controller properties in co-simulation

Feature	Property in co-simulation
Computing speed	The delay time after the execution of the controller agent
Communication speed	The delay time after the execution of the controller agent
Sampling rate	Simulation interval time of power simulation software
Signal–noise ratio	Sampling data from PowerFactory adding specific noise signal

It should be noted that, although the simulation features of the computing speed and the communication speed are reflected in the same delay time after the task execution, the reasons and treatments of them are different. If the computing speed is slow, the time delay of controller will affect the control performance or even lead to control failure. The improvement is to replace with a higher-performance processor or adjust the algorithm. The low communication speed of key nodes will cause congestion of messages. Re-design of the routing path can solve the problem.

9.2.3 Negotiation Between Controllers

In a distributed control system or multi-agent system, the system administrator usually controls the system by setting target rather than control devices directly. The detailed control, such as opening and closing the breaker, is decided by a set of controllers that can communicate, bargain, and negotiate with each other. By using the communication functions provided by the co-simulation environment, agents can communicate with each other in the following two ways:

(1) Negotiation based on protocol. Controllers transfer data, send request, and confirm the conditions via protocols. This communication type is in common with current communication type of power system, so it can be easily to using IEC61850 and other existing protocols.
(2) Negotiation based on semantics. The controller can communicate with each other with dialogues just like we have done every day. This feature is implemented by the FIPA-ACL semantics technology [20]. By semantic analysis, the controller can judge messages more intelligent and act more flexible rather than limited by the protocol. Meanwhile, such a controller can adapt to new types of messages but does not need to modify the code; this makes the controller to have broader compatibility.

9.3 Case Studies

9.3.1 A Distributed Protection Algorithm Based on Local Outlier Factor

Distributed generation and electric vehicles have brought many problems to distribution network, such as multi-directional flows and rapid changes in power. The original fault location and protection methods are no longer valid to the smart grid. Compared with traditional methods, the distributed method of protection can accurately and quickly locate and isolate faults in complex distribution network environment.

A distributed protection algorithm based on local outlier factor (LOF) for smart distribution networks is applied and embedded in the co-simulation of this chapter. This new method introduced the local outlier factor (LOF) algorithm, which is used in the fields of data mining, into the fault location of distribution network, and the better performance was obtained [21].

The measured data of different bus bars in the distribution network has many common features. If we combine the features of a bus bar to a feature matrix, we can get a group matrix representing the feeder line. By analyzing these matrices in a high-dimensional coordinate system, we can observe that the space points representing the feature matrices are gathering together when the grid in normal condition, and the space points representing the abnormal matrices will leave the other points far away in short-circuit condition. This means that, if we can find the outlier in the space points, we can detect the fault location as well. Figure 9.5 shows the flowchart of the proposed fault detection scheme. More specifically, this scheme has the following five steps:

Step 1: Node agent extracts features from the sampling data and sends the information to the zone agent according to a time interval.
Step 2: Zone agent calculates the distance matrix based on the node feature matrix.
Step 3: The LOF algorithm is used to calculate the outlier factor of each node in the distance matrix. According to the features of LOF algorithm, the outliers are far away beyond 1, and all non-outliers are below 1 [20]. Therefore, we can identify the outliers by judging if there is a LOF value more than 1.
Step 4: When one short-circuit fault occurred, the sampling data of switches is internally relative. For example, the voltage RMS, current RMS, and negative-sequence value will change simultaneous. However, this relevance does not exist when the failure is caused by abnormal data or sensor failure. In this step, we use the fault correlation check to distinguish if it is a physical fault or a communication failure. By comparing the characteristic values with fault vectors, we can get the proportion of fault correlation (PFC). If the PFC is beyond the threshold, it means there is no relative in the sampling data; then, it can be judged that it is a communication failure. Otherwise, it is a physical actual fault.
Step 5: Locating the fault position and sending action messages to the corresponding switches. Then the switches open and isolate the fault current.

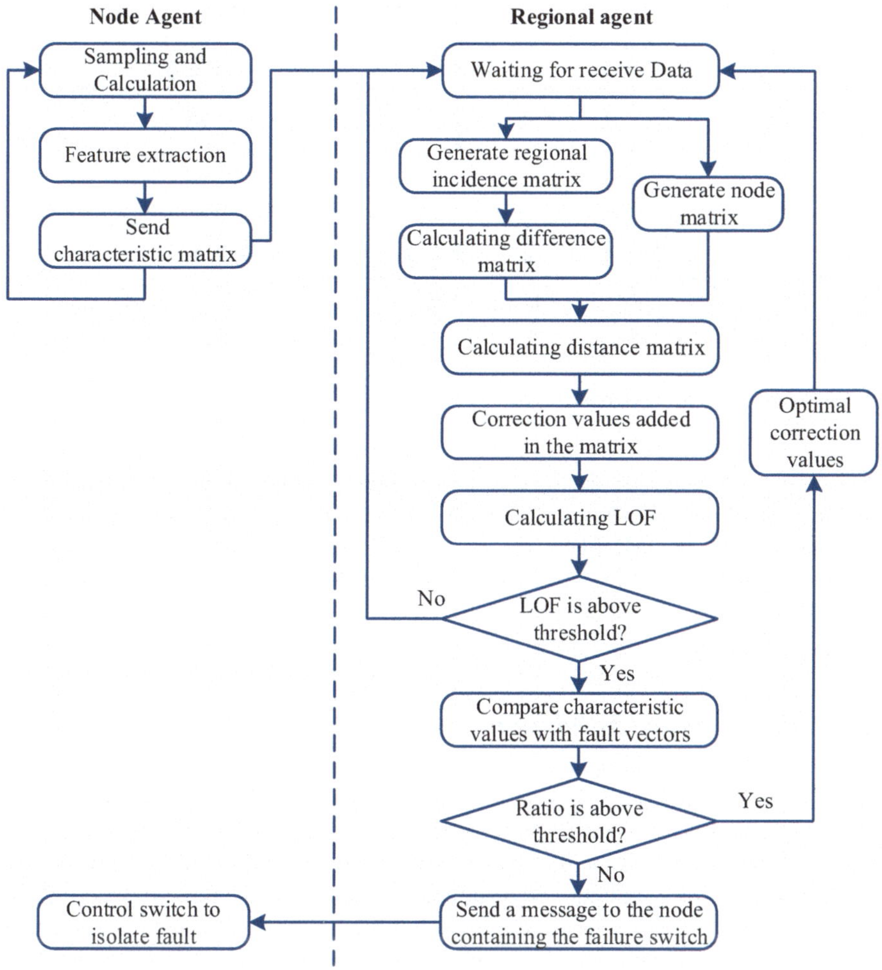

Fig. 9.5 Flowchart of the proposed fault detection scheme

9.3.2 Description of Co-simulation

In order to investigate the distributed fault location and protection algorithm based on LOF, we build a distribution network model in the co-simulation environment, according to one active distribution network in Jiangxi Province, China, as shown in Fig. 9.6.

In Fig. 9.6, the data on the line labels the line power flow, the data near loads and DG are the consumed or generated power, and the entire data unit is p.u. There are two DG accessed to this network at BB3 and BB7, respectively. In this network, the three tie lines, which are connected to the substations located at BB1, BB4, and BB9, respectively, can ensure that the load scheduling is more flexible, but when DG joins,

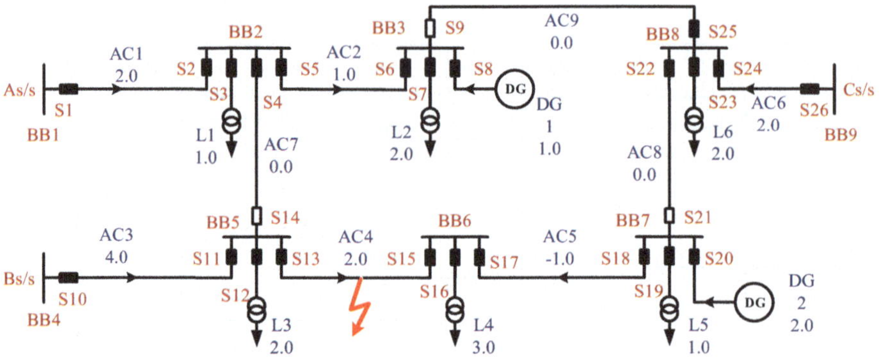

Fig. 9.6 Distribution network model

the operation of the entire network becomes more complex. Figure 9.7 shows the simulation model established in DIgSIENT/PowerFactory and the communication model established in OMNeT++, respectively. These two models are connected by the JADE framework and controlled by the controller agents in the co-simulation environment.

9.3.3 *Performance Validation*

The short-circuit fault, occurred on line AC4, is considered in the case study. Due to the DG2, the fault current flows from two directions to the fault point. In order to test the performance of the distributed fault location and protection algorithm on the identification of communication failure, the sampling of the current on S2 is set abnormally. Figure 9.8 shows the results of the LOF and PFC values. From the results, it is clear that the points including S2, S13, and S15 are outliers. But the PFC of S2 is 100%; it is impossible for a failure that has only one relative characteristic value. The PFC of S13 and S15 is below the threshold, so the area involved S13 and S15 can be identified as the fault area.

When the fault area has been located, the switches S13 and S15 open to isolate the fault; then the actions of power supply recovery and load transfer are excited. After the calculations by agents and the message delay by OMNeT++, all of the actions can be represented in the PowerFactory.

Figure 9.9 shows the negative-sequence currents during fault process. It can be seen that the proposed protection scheme can accomplish fault location and isolation instantaneously, when the range of protection area (RPA) is less than or equal to 20. However, as shown in Table 9.2 about the time delays of router, the network delay occurs and the time of protection process increases substantially when the number of nodes is 21, and it cannot be ignored when RPA reaches to 26. The simulation results reveal the effects of communication network on the protection action in distribution

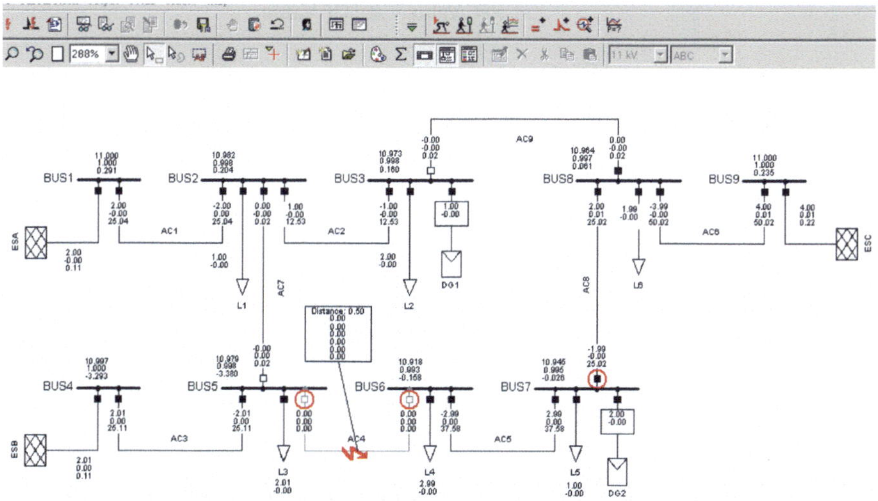

(a) Distribution network model

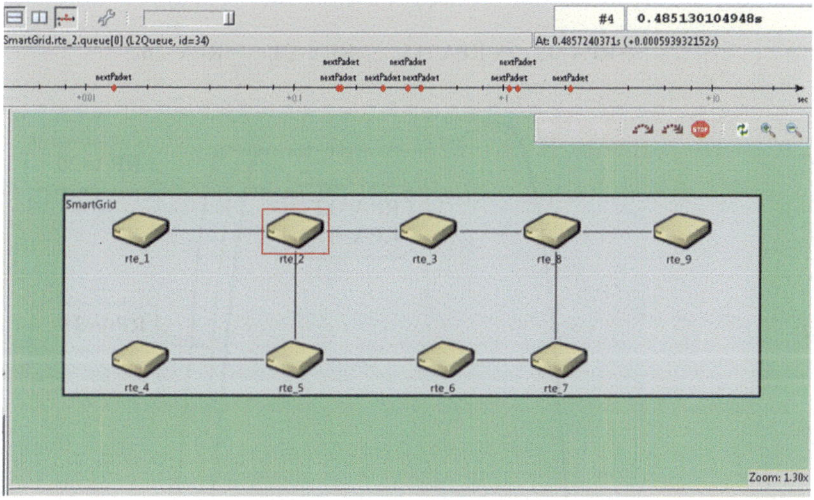

(b) Communication model

Fig. 9.7 Physical model and communication model for co-simulation

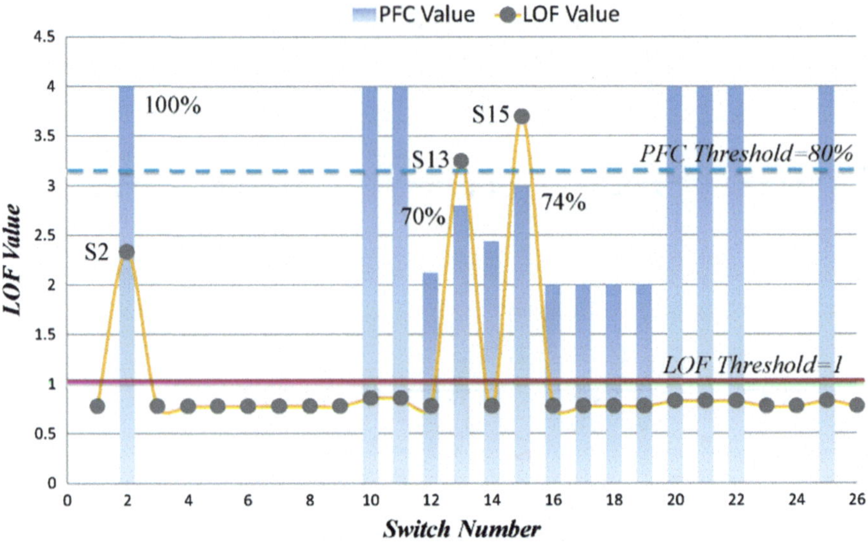

Fig. 9.8 LOF and PFC values when fault happened on line AC4

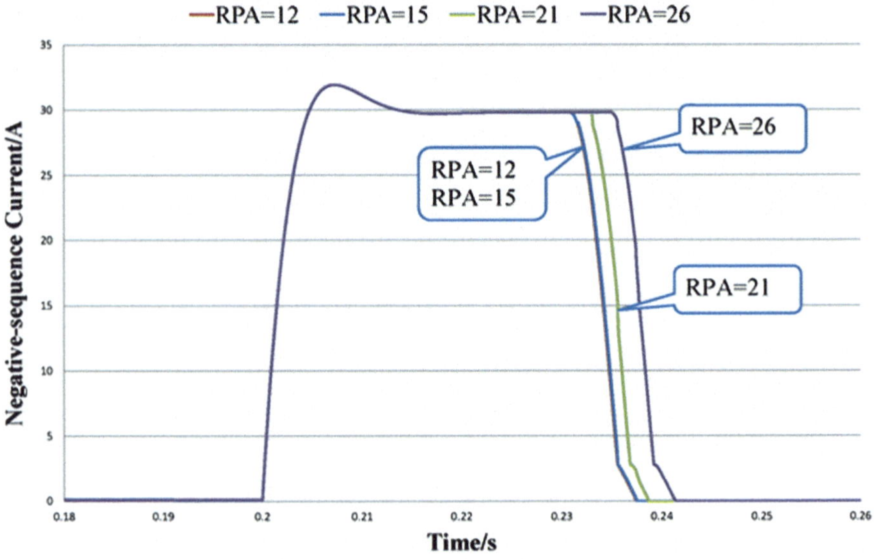

Fig. 9.9 Comparison on negative-sequence currents

Table 9.2 Time delay of communication

Range of protection area	Router delay/ms
12 switches	1.537
17 switches	1.571
21 switches	3.467
26 switches	5.164

networks. Thus, the co-simulation environment is an effective tool to analyze and find the reason of system problems.

9.4 Summary

This chapter proposed a co-simulation environment for protection and control of smart distribution network. This environment can realize the connection of the software PowerFactory and OMNeT++ and is very suitable for the co-simulation of distributed system and multi-agent system. By comparing the performance of the current co-simulation environments, the characteristics of environment, which are fit for the distributed control system, are obtained. According to these characteristics, a co-simulation environment based on JADE framework is proposed. The design method of the distributed controllers is described. The comprehensive evaluation of protection algorithms is carried out via the co-simulation, and the effectiveness of the proposed co-simulation environment is validated.

References

1. Huang AQ, Crow ML, Heydt GT, Zhang JP, Dale SJ (2011) The future renewable electric energy delivery and management (FREEDM) system: the energy internet. Proc IEEE 99(1):133–148
2. Safdarian A, Fotuhi-Firuzabad M, Lehtonen M (2014) A distributed algorithm for managing residential demand response in smart grids. IEEE Trans Ind Inform 10(4):2385–2393
3. Hopkinson K, Wang XR, Giovanini R, Thorp J, Birman K, Coury D (2006) EPOCHS: a platform for agent-based electric power and communication simulation built from commercial off-the-shelf components. IEEE Trans Power Syst 21(2):548–558
4. Ross KJ, Hopkinson KM, Pachter M (2013) Using a distributed agent-based communication enabled special protection system to enhance smart grid security. IEEE Trans Smart Grid 4(2):1216–1224
5. Li W, Monti A, Luo M, Dougal RA (2011) VPNET: a co-simulation framework for analyzing communication channel effects on power systems. In: IEEE electric ship technologies symposium, Alexandria, VA, April 2011
6. Nutaro J, Kuruganti PT, Miller L, Mullen S, Shankar M (2007) Integrated hybrid-simulation of electric power and communications systems. In: IEEE power engineering society general meeting, Tampa, FL, June 2007
7. Liberatore V, Al-Hammouri A (2011) Smart grid communication and co-simulation. In: IEEE Energytech, Cleveland, OH, May 2011

8. Lin H, Veda SS, Shukla SS, Mili L, Thorp J (2012) GECO: global event-driven co-simulation framework for interconnected power system and communication network. IEEE Trans Smart Grid 3(3):1444–1456
9. Chabukswar R, Sinopoli B, Karsai G, Giani A, Neema H, Davis A (2010) Simulation of network attacks on SCADA systems. In: First workshop on secure control systems, cyber physical systems week, Stockholm, Sweden, April 2010
10. Riley D, Eyisi E, Bai J, Koutsoukos X, Xue Y, Sztipanovits J (2011) Networked control system wind tunnel (NCSWT): an evaluation tool for networked multiagent systems. In: International ICST conference on simulation tools and techniques, Brussels, Belgium
11. Georg H, Müller SC, Dorsch N, Rehtanz C, Wietfeld C (2013) INSPIRE: integrated co-simulation of power and ICT systems for real-time evaluation. In: Proceedings of 4th IEEE international conference on smart grid communications, Vancouver, BC, Oct 2013
12. Georg H, Müller SC, Rehtanz C, Wietfeld C (2014) Analyzing cyber-physical energy systems: the INSPIRE Co-simulation of power and ICT systems using HLA. IEEE Trans Ind Inform 10(4):2364–2373
13. Queiroz C, Mahmood A, Tari Z (2011) SCADASim—a framework for building SCADA simulations. IEEE Trans Smart Grid 2(4):589–597
14. SPARKS (2016) Deliverable 2.4 smart grid cyber security simulation environment
15. Mallouhi M, Al-Nashif Y, Cox D, Chadaga T, Hariri S (2011) A testbed for analyzing security of SCADA control systems (TASSCS). In: IEEE PES innovative smart grid technologies, Hilton Anaheim, CA, Jan 2011
16. Duan Y, Luo L, Li Y, Cao Y, Rehtanz C, Küch M (2017) Co-simulation of distributed control system based on JADE for smart distribution networks with distributed generations. IET Gen Transm Distrib 11(12):3097–3105
17. Gonzalez-Longatt F, Rueda JL (2014) PowerFactory applications for power system analysis. Springer (Dec)
18. Wehrle K, Gunes M, Gross J (2010) Modeling and tools for network simulation. Springer (Sept)
19. Bellifemine F, Caire G, Greenwood D (2007) Developing multi-agent systems with JADE. Wiley Series in Agent Technology (Feb)
20. Louis V, Martinez T (2005) An operational model for the FIPA-ACL semantics. In: International workshops on agent communication, Utrecht, Netherlands, July 2005
21. Breunig MM, Kriegel HP, Ng RT (2000) LOF: identifying density-based local outliers. In: Proceedings of ACM SIGMOD2000 international conference on management of data, Dalles, TX

Chapter 10
Local False Data Injection Attacks with Incomplete Network Information

Modern power system is not just a purely independent physical infrastructure incorporating generators, transmission lines, and other electrical units. It is also faced with a growing number of cyber-attacks due to its high dependence on advanced communication system and more open operational environment [1]. When modern power system is attacked by a hacker, it will cause a huge harm to the entire system such as out service of communication system and the mask of overloading lines and generators, and the whole power system could become instable in some extreme cases. A typical example of cyber-attack in power systems is the famous collapse of the Ukraine power system due to cyber-attack [2]. Therefore, it is of vital importance to investigate the cyber-attack on power systems and how to defense smart grid against cyber-attack. Cyber-attack can be mainly divided into the following two types: false data injection attacks and jamming attacks [3].

The literature [4] has shown that the false data c can avoid being detected if it satisfies $c = Ha$. In power systems, H matrix is determined by system topology, line impedances, and also the measurement placement of the system. A common strong assumption in previous researches is that the attacker must have complete knowledge about the power grid topology and transmission line admittances. However, the topological information is confidential in the control center and not easy to obtain. For an attacker with limited attacking ability, it is difficult to have access to the admittance values of all the lines. This chapter will explore the possibility of constructing an undetectable attack vector without knowing the full network information of a power grid.

10.1 False Data Injection for State Estimation

In cyber-physical power system, a hacker could inject false data into the power system by attacking the SCADA system or the measurement device. Taking the SCADA system for example, the hacker may intercept and modify the data transmitted by SCADA to the control center through the fiber eavesdropping technology. This could

Y. Cao et al., *Cyber-Physical Energy and Power Systems*,
https://doi.org/10.1007/978-981-15-0062-6_10

adversely interfere with the power system state estimation and then cause the control center to issue an erroneous command. With the erroneous command, the resulting power system state would be deviated from the scheduled one and the security issues such as overloading of a branch and low voltage of a bus that could be masked. Note that a good hacker should be equipped with so well-specialized power system knowledge that the false data injection can be carefully designed to ensure the cyber-attack on the state estimation that cannot be detected by the bad data detector module [5].

10.1.1 State Estimation of Power System

State estimation is a mathematical process of estimating the real-time state of a power system based on a large number of redundant measurements collected by installed sensors. As the eyes of a power system, the accurate and efficient functioning of state estimation plays a critical role in the stability of electric power systems. For example, inoperative state estimator, which was caused by incorrect telemetry data, started the 2003 Northeastern blackout.

The basic approach to state estimation is called the weighted least square (WLS) [6]. The mathematical model of WLS state estimation is based on the mathematical relations between the measurements and the state variables. The WLS approach is a standard approach to the approximate solution of overdetermined systems, i.e., sets of equations in which there are more equations than unknown states.

The relationship between the measurement vector z (such as bus power injections, line flows, line currents, bus voltages) and state variable vector s (bus voltage magnitudes and angles) is described by

$$z = h(s) + e \tag{10.1}$$

where e represents the vector of measurement errors.

The goal of the state estimation is to determine the best s which minimizes the residual error

$$r = z - h(\hat{s}) \tag{10.2}$$

To achieve the goal, the following optimization problem is solved:

$$\min_{s} J(s) = [z - h(s)]^{\mathrm{T}} W [z - h(s)] \tag{10.3}$$

The optimal s must satisfy

$$\frac{\partial J(s)}{\partial s} = 0 \tag{10.4}$$

which gives

$$H^{\mathrm{T}}(\hat{s})W[z - h(\hat{s})] = 0 \tag{10.5}$$

where

$$H(s) = \frac{\partial h(s)}{\partial s} \tag{10.6}$$

is the Jacobian matrix with respect to state variables.

Equation (10.5) is nonlinear, and its solution can be obtained by an iterative approach.

$$s_{k+1} = s_k + \triangle\, s_k \tag{10.7}$$

The increment $\triangle\, s_k$ is given by (10.8)

$$[G(s_k)] \triangle\, s_k = H^{\mathrm{T}}(s_k)W[z - h(s_k)] \tag{10.8}$$

where

$$G(s) = H^{\mathrm{T}}(s)WH(s) \tag{10.9}$$

The whole algorithm of the state estimation can be summarized as follows:

Step 1: Initialize the state variables s_0 and $k = 0$.
Step 2: Calculate $G(s_k)$, $h(s_k)$.
Step 3: Calculate $\triangle\, s_k$ according to (10.8).
Step 4: Obtain a new state vector.
Step 5: If $|\triangle\, s_k| \leq \varepsilon$, then stop; otherwise $k = k+1$, go to *Step* 1.

There are two methods to determine the initial state variables. The first one is called flat start. The second method is to use a feasible power solution as the initial state.

10.1.2 Complex Network Background

In recent years, cyber security in power systems has attracted intensive research interest. In this chapter, we still consider the direct current (DC) power flow model, in which the voltages at buses $\mathrm{vol}_b \approx 1.0$ p.u. and $\sin\theta \approx \theta$, $\cos\theta \approx 1$. Then, the power injection vector P^{inj} and phase angle vector θ satisfy

$$P^{\mathrm{inj}} = B\theta \tag{10.10}$$

where matrix B is the dependency matrix

$$B = WX^{-1}W^{\mathrm{T}} \tag{10.11}$$

And the line flow is calculated as

$$F = X^{-1}W^{\mathrm{T}}\theta \tag{10.12}$$

Constraints (10.10–10.12) can be treated as a more compact form

$$\begin{bmatrix} P^{\mathrm{inj}} \\ F \\ -F \end{bmatrix} = \begin{bmatrix} WX^{-1}W^{\mathrm{T}} \\ X^{-1}W^{\mathrm{T}} \\ -X^{-1}W^{\mathrm{T}} \end{bmatrix} \theta = H\theta \tag{10.13}$$

where H is defined as the Jacobian matrix of the power grid.

The first consideration in [4] is that if an attacker can construct such a false data injection attack that

$$a = HC \tag{10.14}$$

then, we can have

$$\begin{aligned} \left\| z - H\hat{\theta}_{\mathrm{bad}} \right\|_2 &= \left\| z + \Delta z - H(\hat{\theta} + \Delta \theta) \right\| \\ &= \left\| z - H\hat{\theta} - (\Delta z - H \Delta \theta) \right\| \\ &= \left\| z - H\hat{\theta} \right\|_2 \end{aligned} \tag{10.15}$$

That is, the residue r will not increase, so false data injection attacks on measurements can bypass the residual test.

Considering the physical characteristics of power systems, to make the attack models more practical, the literature [7] introduced the following two constraints into the general false data attack model: (1) Generator output measurements cannot be attacked, i.e., $\Delta G = 0$. Note that this assumption was mainly for large-sized centralized generators. The output measurements for increasing small-sized distributed generators, which can be treated as negative loads, might be attacked. (2) Bus injection measurements of zero-injection buses in the power grid cannot be attacked. (3) Load measurements are attackable, i.e., $|\Delta D \geq 0|$, within certain ranges.

With the above assumptions, the false data injection model is transformed into a load redistribution model (10.16–10.18).

$$\sum_{d=1}^{ND} \Delta D_d = 0 \tag{10.16}$$

$$-\tau D_d \leq \Delta D_d \leq \tau D_d \tag{10.17}$$

$$\Delta F = -S \cdot V \cdot \Delta D \tag{10.18}$$

Constraint (10.16) is true since power balance equation must be met. Constraint (10.17) ensures that injected false data at each bus is within certain ranges. Note that ΔD_d is the false data injection into load measurement d, and τ is the maximum percentage of change for load measurement attack ($0 < \tau < 1$). Constraint (10.18) is true since $\Delta G = 0$. In a load redistribution attack, the attacker aims to modify the load measurements by injecting a false data vector ΔD while keeping the sum of all the false data injection equal to zero. It should be pointed out that load redistribution attacks only redistribute the load readings that come from meters, rather than actually alter physical loads at buses. Load redistribution attack is a special case of false data injection attacks, which captures certain physical characteristics of power systems. However, the attacker has to obtain the full network topology and parameter information and attack a number of measurements to complete such an attack.

10.2 Modeling of Local Data Attacks

10.2.1 Related Work

Several works have been trying to relax the strong condition that the full network information of a power grid is indispensable to constructing an attack vector.

In literature [8], incomplete information was used to model false data injection attacks. In their model, the power grid is divided into two disconnected subnetworks G_1 and G_2 by a cut CU. The set A_1 contains all the non-boundary buses in G_1, and set N_1 contains all the non-boundary buses in G_2. If the attacker knows the admittance values of the lines in the cut and selects specific false data injection such that all the state variables in G_1 and G_2 change the same value, respectively, then the false data would not increase the residue value in the bad data test. In the DC state estimation, state variable refers to bus phase angle. Applying the proposed scheme in [8] to the load redistribution attacking model, the changes in line flows of G_1 and G_2 are all zero since all the buses increase/decrease the same phase angle. Thus, according to KCL law, the injection powers at all buses in A_1 and N_2 will not change. Therefore, there is no injection of false data in A_1 and N_2, and the attacking region is limited in the cut CU. If the attacking targets were beyond the cut, the attacking strategy would fail. Moreover, when all the boundary buses are zero-injection buses, no feasible solution can be found since no attack can be initiated within the cut without being detected.

10.2.2 New Modeling of False Data Injection Attacks

In this section, we propose local load redistribution attacks based on incomplete network information. We assume that the attacker aims to attack region A as shown in Fig. 10.1 [9].

The attacking region is determined by the attacker and is not limited to a cut as in literature [8]. To launch a successful local load redistribution attack, the best strategy is to ensure that the false data injection of powers in the attacking region does not change the power flows in the non-attacking region. This can be done by injecting pre-designed false data making sure phase angle changes are the same for all buses in the non-attacking region and all boundary buses in the attacking region. In fact, we observe that if all boundary buses in the attacking region have the same incremental phase angle, then all buses in the non-attacking region automatically have the same incremental phase angle, thus the power flows in the non-attacking region will not change. This observation is formalized into Theorem 10.1 and mathematically proved in this chapter. In a local load redistribution attack, the attacker does not need to know the complete information of the entire power grid. All the information that the attacker must know are as follows:

- The line incidence matrix of attacking region A, which represents the network topology of the attacking region.
- The admittances of lines in attacking region A, which represents the network parameters of the attacking region.

Theorem 10.1 *Suppose a power grid is decomposed into two connected regions A and N by a set of lines (tie lines). If an additional injected power $\triangle P_{\mathrm{A}}$ into region A makes the phase angles of all its boundary buses increase or decrease the same*

$$\triangle\,\theta_r = \alpha \quad \forall r \in \Omega_{\mathrm{BA}} \tag{10.19}$$

then

(a) *All buses in region N have the same incremental phase angle.*

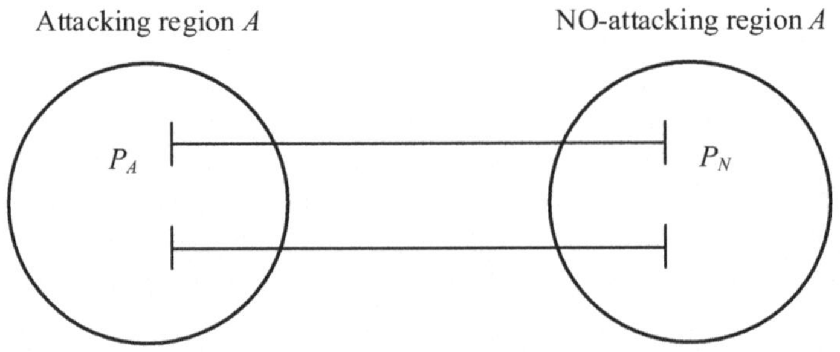

Fig. 10.1 Illustrative diagram for attacking region and non-attacking region

$$\Delta\,\theta_s = \alpha \quad \forall r \in \Omega_N \tag{10.20}$$

(b) *The power flows in region N remain the same.*

$$\Delta F_N = 0 \tag{10.21}$$

(c) *The incremental bus power injection vector and the incremental phase angle vector in region A satisfy*

$$\Delta P_A = B_{A'} \Delta\theta_A \tag{10.22}$$

Proof In (10.23), the B matrix of the entire power grid is reordered such that the rows and columns of all the elements in each submatrix correspond to three regions: Region E contains all the buses in the attacking region except the boundary buses; region C contains all the boundary buses in the attacking region; Region N contains all the buses in the non-attacking region. Note that the attacking region A includes region E and region C. Note also that B_E, B_C, and B_N all include tie lines.

$$\begin{bmatrix} B_E & B_{EC} & 0 \\ B_{CE} & B_C & B_{CN} \\ 0 & B_{NC} & B_N \end{bmatrix} \begin{bmatrix} \Delta\theta_E \\ \Delta\theta_C \\ \Delta\theta_N \end{bmatrix} = \begin{bmatrix} \Delta P_E \\ \Delta P_C \\ \Delta P_N \end{bmatrix} \tag{10.23}$$

From (10.23), we have

$$B_E \Delta\theta_E + B_{EC} \Delta\theta_C = \Delta P_E \tag{10.24}$$

$$B_{CE} \Delta\theta_E + B_C \Delta\theta_C + B_{CN} \Delta\theta_N = \Delta P_C \tag{10.25}$$

$$B_{NC} \Delta\theta_C + B_N \Delta\theta_N = 0 \tag{10.26}$$

Note that (10.26) is true since $\Delta P_N = 0$ for the non-attacking region N. We choose one boundary bus in region C as the reference bus and let its incremental phase angle be α. Then, according to the conditions in Theorem 10.1, the incremental phase angles of all boundary buses in region C are α,

$$\Delta\theta_C = \alpha_C \tag{10.27}$$

In DC power flow, the phase angles of all buses can be increased by the same amount without affecting the power flow solutions. Thus,

$$B_E(\Delta\theta_E - \alpha_E) + B_{EC}(\Delta\theta_C - \alpha_C) = \Delta P_E \tag{10.28}$$

$$B_{CE}(\Delta\theta_E - \alpha_E) + B_C(\Delta\theta_C - \alpha_C) + B_{CN}(\Delta\theta_N - \alpha_N) = \Delta P_C \tag{10.29}$$

$$B_{NC}(\Delta\theta_C - \alpha_C) + B_N(\Delta\theta_N - \alpha_N) = 0 \tag{10.30}$$

Substituting (10.27) into (10.28–10.30), we obtain

$$B_E(\Delta\theta_E - \alpha_E) = \Delta P_E \tag{10.31}$$

$$B_{CE}(\Delta\theta_E - \alpha_E) + B_{CN}(\Delta\theta_N - \alpha_N) = \Delta P_C \tag{10.32}$$

$$B_N(\Delta\theta_N - \alpha_N) = 0 \tag{10.33}$$

Note that B_N contains the tie lines, thus it is the bus dependency matrix of the sub-network which can be treated as a ground-connected network; therefore, the inverse of B_N exists, whether the non-attacking region N is connected or not. Therefore, from (10.33), we have

$$\Delta\theta_N = \alpha_N \tag{10.34}$$

Accordingly,

$$\Delta F_N = X_N^{-1} W_N^{\mathrm{T}} \Delta\theta_N = 0 \tag{10.35}$$

Substituting (10.34) into (10.32), we have

$$B_{CE}(\Delta\theta_E - \alpha_E) = \Delta P_C \tag{10.36}$$

As defined in Theorem 10.1, $B_{A'}$ is the bus dependency matrix in region A ignoring the tie lines. $B_{A'}$ can also be rewritten as

$$B_{A'} = \begin{bmatrix} B_E & B_{EC} \\ B_{CE} & B_{C'} \end{bmatrix} \tag{10.37}$$

Note that $B_{C'}$ is the bus susceptance matrix in region C ignoring the tie lines. Since $B_{A'}$ does not include the tie lines, the summation of elements of any row or any column in $B_{A'}$ is zero. So, $B_{A'}$ is singular, and we have

$$\begin{bmatrix} B_E & B_{EC} \\ B_{CE} & B_{C'} \end{bmatrix} \begin{bmatrix} \alpha_E \\ \alpha_C \end{bmatrix} = 0 \tag{10.38}$$

We also have

$$\Delta\theta_A = \begin{bmatrix} \Delta\theta_E \\ \Delta\theta_C \end{bmatrix} \tag{10.39}$$

$$\Delta P_A = \begin{bmatrix} \Delta P_E \\ \Delta P_C \end{bmatrix} \tag{10.40}$$

Therefore,

$$\begin{aligned} B_{A'}\Delta\theta_A &= \begin{bmatrix} B_E & B_{EC} \\ B_{CE} & B_{C'} \end{bmatrix}\begin{bmatrix} \Delta\theta_E \\ \Delta\theta_C \end{bmatrix} \\ &= \begin{bmatrix} B_E & B_{EC} \\ B_{CE} & B_{C'} \end{bmatrix}\begin{bmatrix} \Delta\theta_E \\ \Delta\theta_C \end{bmatrix} - \begin{bmatrix} B_E & B_{EC} \\ B_{CE} & B_{C'} \end{bmatrix}\begin{bmatrix} \alpha_E \\ \alpha_C \end{bmatrix} \\ &= \begin{bmatrix} B_E & B_{EC} \\ B_{CE} & B_{C'} \end{bmatrix}\begin{bmatrix} \Delta\theta_E - \alpha_E \\ \Delta\theta_C - \alpha_E \end{bmatrix} \end{aligned} \tag{10.41}$$

Based on (10.30), (10.31), and (10.36), then we have

$$\begin{aligned} B_{A'}\Delta\theta_A &= \begin{bmatrix} B_E & B_{EC} \\ B_{CE} & B_{C'} \end{bmatrix}\begin{bmatrix} \Delta\theta_E - \alpha_E \\ \Delta\theta_C - \alpha_C \end{bmatrix} \\ &= \begin{bmatrix} B_E & B_{EC} \\ B_{CE} & B_{C'} \end{bmatrix}\begin{bmatrix} \Delta\theta_{\mathrm{E}} - \alpha_E \\ 0 \end{bmatrix} \\ &= \begin{bmatrix} \Delta P_E \\ \Delta P_C \end{bmatrix} \\ &= \Delta P_A \end{aligned} \tag{10.42}$$

Thus, Theorem 10.1 is proved based on (10.34), (10.35), and (10.42).

In this chapter, we use Theorem 10.1 to analyze local load redistribution attacks. Note that both the attacking and non-attacking regions are connected as shown in Fig. 10.1.

$$\Delta P_N = 0 \tag{10.43}$$

$$\Delta G_A = 0 \tag{10.44}$$

Constraint (10.43) is true since region N is the non-attacking region. Constraint (10.44) is true since generator measurements cannot be attacked in load redistribution attacks.

According to Theorem 10.1, if the attacker designs the false data vector ΔD such that the variations of phase angles of the boundary buses in region A are the same (e.g., β), then we have

$$\Delta\theta_r = \beta \qquad \forall r \in \Omega_{BA} \cup \Omega_N \tag{10.45}$$

Since generator output measurements cannot be attacked, i.e., $\Delta G_A = 0$, we have

$$\Delta P_A = U_A \Delta G_A - V_A \Delta D_A = -V_A \Delta D_A \tag{10.46}$$

Substituting (10.46) into (10.22), we obtain

$$B_{A'} \Delta\theta_A = \Delta P_A = -V_A \Delta D_A \tag{10.47}$$

To sum up, if the false data injection vector ΔD_A ensures that the phase angles of all boundary buses in the attacking region A increase or decrease the same, then the false data injection would not impact the power flows in the non-attacking region. Since load measurements are attackable within certain ranges ($0 < \tau < 1$), we ensure that (10.48) is valid, and the line flow injection vector in region A is constructed by (10.49).

$$-\tau \mathrm{D}_{A,d} \leq \Delta \mathrm{D}_{A,d} \leq \tau \mathrm{D}_{A,d} \tag{10.48}$$

$$\Delta F_A = X_A^{-1} W_A^{\mathrm{T}} \Delta\theta_A \tag{10.49}$$

Thus, any arbitrary nonzero false injection power vector $[\Delta D_A, \Delta F_A]$, which satisfies (10.19), (10.47), (10.48), and (10.49) can be used to launch a successful local load redistribution attack. In other words, (10.19), (10.47), (10.48), and (10.49) together define the conditions an attacking vector has to satisfy. However, they do not tell how to select a feasible attacking region so that those conditions can be satisfied. We will address the problem in the next chapter.

10.3 Impacts of Network Connectivity

The connectivity of a network has an impact on the construction of an attacker vector. In this section, the connective of both the attacking region and non-attacking region is considered. We have discussed the case where both the attacking region and non-attacking region are connected. In fact, both the attacking region and non-attacking region could be disconnected. In this section, we will further investigate how the connectivity would affect the phase angles of boundary buses, and accordingly the attacking model.

It is also important to study the cases where the attacking and/or non-attacking regions are disconnected. For example, an attacker may attack a set of meters installed at buses with relatively heavy loads so that it will lead to more serious damage. If these buses were not clustered in one connected subnetwork, then the attacking region would be disconnected.

10.3.1 Disconnection Case 1

The attacking region A is disconnected, and the non-attacking region N is connected. Specifically, the attacking region A is composed of m islands $A_1, A_2, \cdots A_m$; each island $A_i (i = 1, \cdots m)$ is connected to the non-attacking region through boundary buses in set Ω_{BA_i}. If the attacker aims to attack the measurements in region A without impacting the measurements in region N, the sufficient condition is that false data injection of powers into each attacking region A_i will not change power flows in the non-attacking region N. For any two different attacking regions A_{i1} and A_{i2}, if the phase angles of the boundary buses in each attacking region increase or decrease by the same (β_{i1} and β_{i2}, respectively), then according to Theorem 10.1, all the buses in the non-attacking region N have the same incremental phase angle as that of the boundary buses in the attacking region A_{i1} and A_{i2}

$$\Delta\theta_S = \beta_{i1}, \quad \forall s \in \Omega_N \tag{10.50}$$

$$\Delta\theta_S = \beta_{i2}, \quad \forall s \in \Omega_N \tag{10.51}$$

Thus, from (10.50) and (10.51), we can have

$$\beta_{i1} = \beta_{i2}, \quad \forall i1 \neq i2 \tag{10.52}$$

That is the phase angles of all boundary buses in the attacking region A increase or decrease the same. Because all the boundary buses and the buses in the non-attacking region have the same incremental phase angle, there would be no power flows out of each attacking region, thus, (10.47) holds for each attacking region.

10.3.2 Disconnection Case 2

The attacking region A is connected, and the non-attacking region N is disconnected. Specifically, the non-attacking region N is composed of n islands $N_1, N_2, \cdots N_n$; the attacking region A is connected to island N_j through boundary buses in set $\Omega_{BA_i} (j = 1, \cdots, n)$. If the attacker designs a false data vector $\Delta D(-\tau D \leq \Delta D \leq \tau D)$ such that the variations of phase angles of boundary buses in the attacking region A satisfy Proposition 10.1 that is

$$\Delta\theta_r = \beta_j, \quad \forall r \in \Omega_{BA_j} \quad (j = 1, 2, \ldots, n) \tag{10.53}$$

Once (10.53) is satisfied, according to Theorem 10.1,

$$\Delta\theta_S = \beta_j, \quad \forall s \in \Omega_{N_j} \quad (j = 1, 2, \ldots, n) \tag{10.54}$$

$$\Delta F_{N_j} = 0 \quad (j = 1, 2, \ldots, n) \tag{10.55}$$

Constraint (10.55) indicates that the power flows in the non-attacking region remain unchanged. Equations (10.53) and (10.54) together indicate that the flows passing through the tie lines are zeros (i.e., there are no additional power flows in or out of the attacking region A), thus (10.47) holds.

10.3.3 Disconnection Case 3

The attacking region A is disconnected, and the non-attacking region N is also disconnected. Specifically, the attacking region A is composed of m islands $A_1, A_2, \cdots A_m$, and the non-attacking region N is composed of n islands $N_1, N_2, \cdots N_m$. Disconnection case 3 is the combination of disconnection case 1 and disconnection case 2, thus the sufficient conditions are as follows: For each attacking region A_i, the boundary buses connected to the same non-attacking region N_j have the same incremental phase angle. Then, (10.47) holds for each attacking region according to Theorem 10.1, and the power flows in the non-attacking region N will not change. So, the attacker can launch a local load redistribution attack.

So far, we have proved the following proposition that covers all cases discussed above:

Proposition 10.1 *Boundary buses in the attacking region connected to the same island must have the same incremental phase angle. However, boundary buses in the attacking region connected to different islands could have different incremental phase angles.*

It is important to note that, if the attacker does not have the knowledge of the topology information of the non-attacking region, i.e., if the attacker does not know whether the non-attacking region is connected or not, a stricter condition would be that all boundary buses in the attacking region have the same incremental phase angle. This essentially means that, in (10.54), all β_j are the same.

10.4 Feasibility of Attack Vectors

The local attacking scheme shows that an attacker can attack a local region without knowing any network information of external non-attacking region. However, we found that the selection of the attacking region is not arbitrary. In particular, for some specific region, we cannot construct a feasible nonzero attacking vector. Thus, it is necessary to find a rule to determine a feasible attacking region.

Theorem 10.2 *Suppose a connected power grid is decomposed into two connected regions A and N by a set of lines (tie lines). Suppose the attacking region A consists* *of ρ non-boundary buses and σ boundary buses. If there are at most $q = \rho - 1$ non-attackable bus injection measurements in region A, then there exists a feasible non-zero attacking vector.*

Theorem 10.2 provides a graph theoretical approach to determine a feasible attacking region. What needs to do is to count the numbers of boundary and non-boundary buses in the attacking region.

Theorem 10.2 can also be extended to include the cases where the attacking region A is disconnected and/or the non-attacking region N is disconnected, as shown in Theorem 10.2E.

Theorem 10.2E *Suppose a connected power grid is decomposed into an attacking region A and a non-attacking region N by a set of lines (tie lines). Suppose the attacking region A consists of ρ non-boundary buses and σ boundary buses. The σ boundary buses in A are connected to n non-attacking islands. If there are at most $q = \rho + n - 2$ non-attackable bus injection measurements in region A, then there exists a feasible non-zero attacking vector.*

Proof of Theorem 10.2E We assume that the attacking region A consists of m islands. For each attacking island $i (i = 1, 2, \ldots, m)$, according to Theorem 10.1 and the discussions above, (10.54) holds, so

$$\Delta P_{A_i} = B_{A'_i} \Delta\theta_{A_i} \tag{10.56}$$

Then, based on (10.56), for the entire attacking region A, we have

$$\begin{bmatrix} \Delta P_{A_1} \\ \Delta P_{A_2} \\ \vdots \\ \Delta P_{Am} \end{bmatrix} = \begin{bmatrix} B_{A'_1} & & & \\ & B_{A'_2} & & \\ & & \ddots & \\ & & & B_{A'_m} \end{bmatrix} \begin{bmatrix} \Delta\theta_{A_1} \\ \Delta\theta_{A_2} \\ \vdots \\ \Delta\theta_{A_m} \end{bmatrix} \tag{10.57}$$

Equation (10.56) can be rewritten in a more compact way as

$$\Delta P_M = B_{M'} \Delta\theta_M \tag{10.58}$$

Next, we investigate how many independent variables the incremental phase angle vector $\Delta\theta_m$ has. Firstly, the ρ non-boundary buses will introduce ρ independent variables into $\Delta\theta_m$. Secondly, according to Proposition 10.1, the boundary buses that are connected to the same non-attacking island have the same incremental phase angle. That is, the incremental phase angles of the buses connected to the same non-attacking island can be represented by one independent variable, so all the boundary buses will only introduce n independent variables (instead of σ) into $\Delta\theta_m$. Therefore, there are $\mu = \rho + n$ independent variables in $\Delta\theta_m$. Let Ω_0 be the set of buses whose bus injection measurements are not attackable and p be the number of elements in Ω_0. Since there are at most $q = \rho + n - 2$ non-attackable bus injection measurements,

$$p \le \rho + n - 2 \tag{10.59}$$

To prove Theorem 10.2E, we only need to show that there always exists one phase angle vector $\Delta\theta_m$ that satisfies (10.59) and (10.60).

$$-\Phi_M \le \Delta P_M = B_{M'}\Delta\theta_M \le \Phi_M \tag{10.60}$$

$$\Delta P_M = B_M \Delta\theta_{M'} \ne 0 \tag{10.61}$$

where $\Phi_{M,l} = 0, \forall l \in \Omega_0$ and $\Phi_{M,l} = \tau|D_l|, \forall l \notin \Omega_0$, and D_l is the load at bus l. Set

$$\Delta P_{M,l} = 0, \quad \forall l \in \Omega_0 \tag{10.62}$$

Since the total number of independent variables $\Delta\theta_m$ and equations in (10.61) is $\mu = \rho + n$ and p, respectively, and $p \le \rho + n - 2$, there exists an infinite solution $\Delta\theta_m$ for (10.61).

Next, we prove that there are at least two elements in $\Delta\theta_m$ with different values. Assuming that all elements in $\Delta\theta_m$ are equal, that is, the solution of (10.61) satisfies

$$\Delta\theta_M = c \begin{bmatrix} 1 \\ 1 \\ \vdots \\ 1 \end{bmatrix} \tag{10.63}$$

which shows that the rank of solution space is 1. Since there are $\mu = \rho + n$ independent variables, we have

$$\text{rank}(B_z) = \rho + n - 1 \tag{10.64}$$

where B_Z is the coefficient matrix of (10.61).

Since there are $p \le \rho + n - 2$ equations in (10.61),

$$\text{rank}(B_z) \le \rho + n - 2 \tag{10.65}$$

which contradicts (10.63). Thus, there are at least two elements in $\Delta\theta_m$ with different values.

Since $\Delta\theta_m$ is composed of m islands, there must be a vector $\Delta\theta_{A_i} \ne 0$ with at least two elements with different values.

Pick one bus as the reference bus and set its incremental phase angle to be zero. Then remove the row and column in (10.56) corresponded to the reference bus, we have the reduced equation

$$\Delta P_{A_i}^R = B_{A_i'}^R \Delta\theta_{A_i}^R \tag{10.66}$$

Since there are at least two elements in $\Delta\theta_{A_i}$ with different values, we can have $\Delta\theta_{A_i}^R \neq 0$.

Note that $B_{A_i'}^R$ is a full rank matrix, and $\Delta\theta_{A_i}^R \neq 0$, so $\Delta P_{A_i}^R \neq 0$ This is because that if $\Delta P_{A_i}^R \neq 0$, then we can have

$$\Delta\theta_{A_i}^R = B_{A_i'}^{R-1} \Delta P_{A_i}^R = 0 \tag{10.67}$$

Contradiction. Hence, (10.60) has been proved.

Using this $\Delta\theta_m$, we can get $\Delta P_{M,k}$, $\forall k \notin \Omega_0$. Define

$$\gamma_k = \begin{cases} 1, & \text{if } \left|\Delta P_{M,K}\right| \leq \Phi_{M,K} \\ \Phi_{M,K}/\left|\Delta P_{M,K}\right|, & \text{if } \left|\Delta P_{M,K}\right| > \Phi_{M,K} \end{cases} \quad \forall k \notin \Omega_0 \tag{10.68}$$

$$\gamma = \min\{\gamma_k, \forall k \notin \Omega_0\} \tag{10.69}$$

Based on (10.67) and (10.68),

$$0 < \gamma \leq 1 \tag{10.70}$$

Let

$$\Delta\theta_M = \gamma \Delta\theta_M \tag{10.71}$$

It can be verified that, with the new $\Delta\theta_m$ in (10.70), (10.59) and (10.60) are satisfied, which proves Theorem 10.2E.

Theorem 10.2 is a special case of Theorem 10.2E with $n = 1$. In essence, if the attacker does not know the topology of the non-attacking region and treats it as one island ($n = 1$), then the condition is relaxed to "if there are at most $\rho - 1$ non-attackable bus injection measurements in region A."

If we want to make sure that a specific attacking island has a feasible nonzero attacking vector, then Theorem 10.2E can be modified as follows:

Theorem 10.2F *Suppose a connected power grid is decomposed into an attacking region A and a non-attacking region N by a set of lines (tie lines). For a specific attacking island in the attacking region A, suppose there are ρ non-boundary buses and σ boundary buses, and the σ boundary buses in the attacking island are connected to n non-attacking islands. If there are at most $q = \rho + n - 2$ non-attackable bus injection measurements in the attacking island, then there exists a feasible non-zero attacking vector in the attacking island.*

10.5 Case Studies

In this section, we test the local load redistribution attack model using the modified IEEE 14-bus system [10]. Lines 2–3, 2–4, 6–13, and 7–9 are removed from the original system for the purpose of illustrating the concepts in this chapter. The system is composed of 14 buses and 20 transmission lines. Simulations are carried out on a 3.4 GHz personal computer with 8 GB of RAM. The model and algorithm are implemented in MATLAB.

We assume that this system is fully measured. That is, we need one meter to measure the injection power for each bus and two meters to measure the power flow passing through each transmission line. Thus, 54 measurements are needed in total. Attack magnitude for a load bus is limited at $\tau = \pm 50\%$ of the actual load.

Note that, in the local load redistribution attacking model, generator output measurements cannot be attacked, and all the measurements in the non-attacking region cannot be attacked either. The attacker aims to launch a local load redistribution attack without being detected by the control center with the following assumptions:

(1) The attacker has full knowledge of the network topology and the network parameters of the attacking region.
(2) The attacker has no knowledge of the network topology and the network parameters of the non-attacking region.
(3) The attacker has the ability to attack all load and line flow measurements in the attacking region.

We consider the four attacking cases proposed above.

Connection Case
The attacking region, circled by the dashed line in Fig. 10.2, is connected and composed of buses 6, 9, 10, 11, 12, 13, 14 and lines 6–11, 6–12, 9–10, 9–14, 10–11,

Fig. 10.2 Connection case

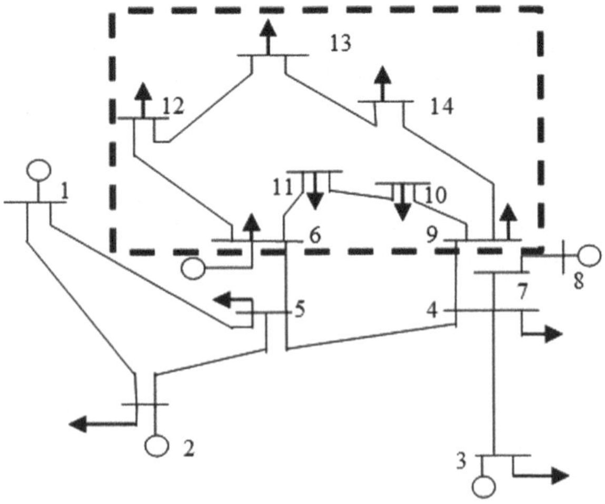

12–13, 13–14. The non-attacking region is also connected if the attacking region is removed. Buses 6 and 9 are the boundary buses in the attacking region. Buses 4 and 5 are the boundary buses in the non-attacking region. According to Proposition 1, a valid and effective local load redistribution attack will make boundary buses 6 and 9 have the same phase angle change. If we choose bus 6 as the reference bus, then we have $\Delta\theta_6 = \Delta\theta_9 = 0$. Note that in this case, there are two non-boundary buses ($\rho = 2$) and no non-attackable bus injection measurement ($q = 0$) in the attacking region. Since $q < \rho - 1$, based on Theorem 10.2, a feasible nonzero attacking vector is guaranteed.

Disconnection Case 1

The attacking region is disconnected and composed of two islands, which are circled by the dashed line and solid line, respectively, in Fig. 10.3. The upper attacking island includes buses 12, 13, 14 and lines 12–13, 13–14, with boundary buses 12 and 14. The lower attacking island includes buses 1, 2, 3, 4, 5 and line 1–2, 1–5, 2–5, 3–4, 4–5, with boundary buses 4 and 5. The non-attacking region, with boundary buses 6 and 9, remains connected if we remove the two attacking islands. According to Proposition 10.1, a valid and effective local load redistribution attack will make all boundary buses in the attacking region have the same phase angle change. If we choose bus 4 as the reference bus, then we have $\Delta\theta_4 = \Delta\theta_5 = \Delta\theta_{12} = \Delta\theta_{14} = 0$. Note that in this case, all the boundary buses in the attacking region are connected to the same non-attacking island ($n = 1$), and there are four non-boundary buses ($\rho = 4$) and one non-attackable bus injection measurement ($q = 1$ for bus 1) in the attacking region. Since $q < \rho + n - 2$, based on Theorem 10.2E, a feasible nonzero attacking vector is guaranteed.

Disconnection Case 2

The attacking region, circled by the dashed line in Fig. 10.4, is connected and includes

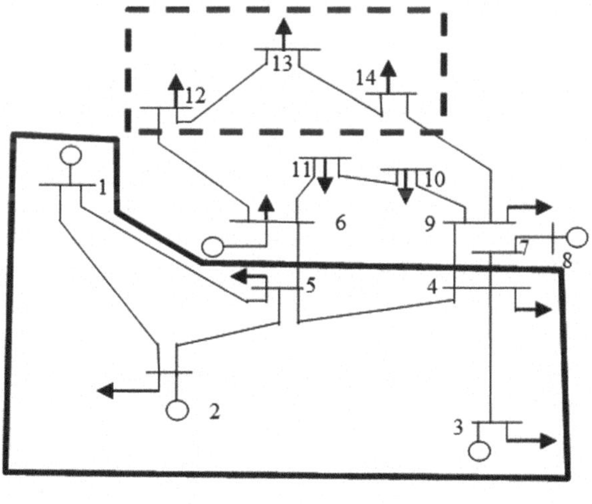

Fig. 10.3 Disconnection case 1

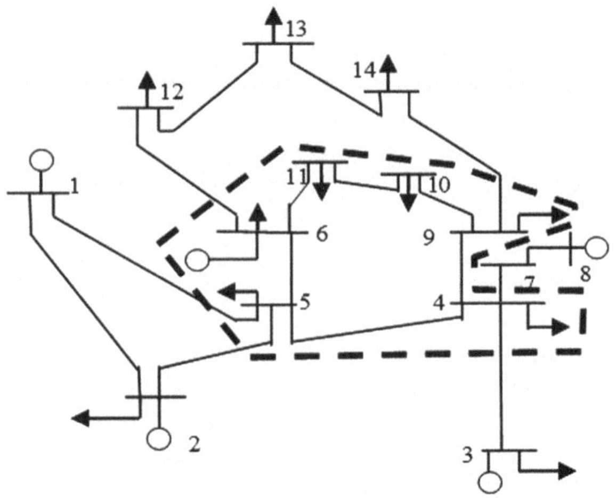

Fig. 10.4 Disconnection case 2

buses 4, 5, 6, 9, 10, 11 and lines 4–5, 4–9, 5–6, 6–11, 9–10, 10–11, with buses 4, 5, 6, and 9 being the boundary buses. The non-attacking region is separated into three islands if we remove the attacking region. Island 1 includes buses 12, 13, 14 and lines 6–12, 9–14, 12–13, 13–14, with buses 12 and 14 being the boundary buses. Island 2 includes buses 1, 2 and lines 1–2, 1–5, 2–5 with buses 1 and 2 being the boundary buses. Island 3 includes buses 3, 7, 8 and lines 3–4, 4–7, 7–8, with buses 3 and 7 being the boundary buses.

Accordingly, boundary buses of the attacking region are divided into three parts. Buses 6 and 9 are connected to the upper Island 1, bus 5 is connected to the left lower Island 2, and bus 4 is connected to the right lower Island 3. If the attacker does not know the topology information of the non-attacking region, the non-attacking region can be treated as a single island. According to Proposition 10.1, boundary buses in the attacking region connected to the same island must have the same incremental phase angle, thus a valid and effective local load redistribution attack will make all boundary buses in the attacking region have the same phase angle change.

If we choose bus 6 as the reference bus, then we have $\Delta\theta_4 = \Delta\theta_5 = \Delta\theta_6 = \Delta\theta_9 = 0$. However, if the attacker knows the topology of the non-attacking region, the condition could be further relaxed. Again, if we choose bus 6 as the reference bus, then we have $\Delta\theta_6 = \Delta\theta_9 = 0$. However, boundary buses 4 and 5 could have a different phase angle change since they are connected to a different non-attacking island. The later condition, assuming that the attacker knows the topology information of the non-attacking region, is less strict than the former condition, assuming that the attacker does not know the topology information of the non-attacking region. Note that in this case, the boundary buses are connected to two non-attacking islands ($n = 2$), and there are two non-boundary buses ($\rho = 2$) and no non-attackable bus injection measurement ($q = 0$). Since $q < \rho + n - 2$, based on Theorem 10.2E, a feasible nonzero attacking vector is guaranteed.

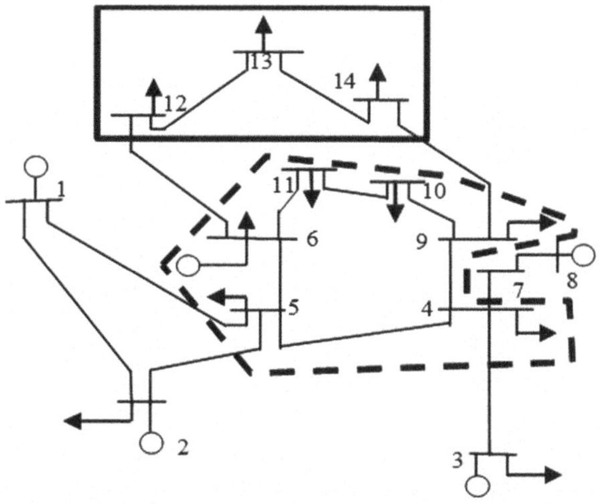

Fig. 10.5 Disconnection case 3

Disconnection Case 3

The attacking region is disconnected and composed of two separate islands, which are circled by the dashed line and solid line, respectively, in Fig. 10.5. The upper attacking Island 1 includes buses 12, 13, 14 and lines 12–13, 13–14, with buses 12 and 14 being the boundary buses. The lower attacking Island 2 includes buses 4, 5, 6, 9, 10, 11 and lines 4–5, 4–9, 5–6, 6–11, 9–10, 10–11, with buses 4, 5, 6, and 9 being the boundary buses.

The non-attacking region is separated into four islands if we remove the attacking region. Island 1 only includes line 6–12. Island 2 only includes line 9–14. Note that Islands 1 and 2 are special in that they only include lines but do not include any bus. Island 3 includes buses 1, 2 and lines 1–2, 1–5, 2–5 with buses 1 and 2 being the boundary buses. Island 4 includes buses 3, 7, 8 and lines 3–4, 4–7, 7–8, with buses 3 and 7 being the boundary buses. Accordingly, boundary buses of the attacking region are divided into four parts. Buses 6 and 12 are connected to the left upper Island 1, buses 9 and 14 are connected to the right upper Island 2, bus 5 is connected to the lower left Island 3, and bus 4 is connected to the lower right Island 4. If the attacker does not know the topology information of the non-attacking region, the non-attacking region can be treated as a single island.

According to Proposition 10.1, boundary buses in the attacking region connected to the same island must have the same incremental phase angle, thus a valid and effective local load redistribution attack will make all boundary buses in the attacking region have the same phase angle change. If we choose bus 6 as the reference bus, then we have $\Delta\theta_4 = \Delta\theta_5 = \Delta\theta_6 = \Delta\theta_9 = \Delta\theta_{12} = \Delta\theta_{14} = 0$. However, if the attacker knows the topology of the non-attacking region, the condition could be further relaxed. Again, according to Proposition 1, if we choose bus 6 as the reference bus, then the constraints for the phase angles of boundary buses are $\Delta\theta_6 = \Delta\theta_{12} = 0$, $\Delta\theta_9 = \Delta\theta_{14}$. The later condition, assuming that the attacker knows the topology information of the non-attacking region, is less strict than the former

Table 10.1 False data into load measurements for connection case

Bus	$\Delta\theta_d$	ΔD_d(MW)	τD_d (MW)
6	0	−0.5134	5.60
9	0	−3.5644	14.75
10	0.0787	0.6810	4.50
11	0.1268	0.8876	1.75
12	−0.0316	0.1687	3.05
13	−0.0901	−2.5968	6.75
14	0.7112	4.9376	7.45

condition, assuming that the attacker does not know the topology information of the non-attacking region. Note that in this case, the boundary buses are connected to four non-attacking islands ($n = 4$), and there are three non-boundary buses ($\rho = 3$) and no non-attackable bus injection measurement ($q = 0$). Since $q < \rho + n - 2$, based on Theorem 10.2E, a feasible nonzero attacking vector is guaranteed.

Tables 10.1, 10.2, 10.3, and 10.4 list the variations of bus phase angles and false data injections at load buses in the attacking region, for the four cases, respectively. In particular, in Tables 10.3 and 10.4, the second and third columns represent incremental bus phase angles and bus injection powers assuming that the attacker has no information of topology and parameters of the non-attacking region, and the fourth and fifth columns assuming that the attacker has the topology information but not parameter information of the non-attacking region. As can been seen, all the false data injections at load buses are summed to zero and limited within $[-\tau D_d, \tau D_d]$, and the false data injections at generator buses are zero. Note that Tables 10.3 and 10.4 show that, if the topology of non-attacking region is unknown to the attacker (columns 2 and 3) and, accordingly, all boundary buses in the attacking region have the same incremental phase angle ($\Delta\theta_4 = \Delta\theta_5 = \Delta\theta_6 = \Delta\theta_9 = 0$), then load buses 4 and 5 in the attacking region become non-attackable buses, and accordingly, $\Delta D_4 = \Delta D_5 = 0$. However, if the topology information of the non-attacking region is known to the attacker (columns 4 and 5), buses 4 and 5 become attackable buses.

Table 10.2 False data into load measurements for disconnection case 1

Bus	$\Delta\theta_d$	ΔD_d(MW)	τD_d (MW)
1	0.1174	0	0
2	0.1485	1.3802	47.10
3	0.1898	1.1097	23.90
4	0	−1.1097	3.80
5	0	−1.3802	5.60
12	0	−2.9767	3.05
13	0.5950	4.6863	6.75
14	0	−1.7096	7.45

Table 10.3 False data into load measurements for disconnection case 2

Bus number	Topology of non-attacking region unknown		Topology of non-attacking region known		τD_d (MW)
	$\Delta\theta_d$	ΔD_d (MW)	$\Delta\theta_d$	ΔD_d (MW)	
4	0	0	−0.1507	−4.0120	23.90
5	0	0	0.0069	3.7684	3.80
6	0	−0.0266	0	−0.0006	5.60
9	0	1.4391	0	−1.1687	14.75
10	0.1216	−2.0997	0.1216	2.1006	4.50
11	−0.0053	0.6873	−0.0053	−0.6877	1.75

Table 10.4 False data into load measurements for disconnection case 3

Bus number	Topology of non-attacking region unknown		Topology of non-attacking region known		τD_d (MW)
	$\Delta\theta_d$	ΔD_d (MW)	$\Delta\theta_d$	ΔD_d (MW)	
4	0	0	0.0659	−3.4443	23.90
5	0	0	−0.0613	3.2639	3.80
6	0	−0.4193	0	−0.6625	5.60
9	0	−1.5716	−0.1697	0.8603	14.75
10	−0.1328	1.8288	−0.1328	−0.1795	4.50
11	−0.0834	0.1621	−0.0834	0.1621	1.75
12	0	1.9732	0	1.9732	3.05
13	0.3944	−3.1065	0.3944	−3.5941	6.75
14	0	1.1333	−0.1697	1.6209	7.45

Next, we test the impact of local load redistribution attacks for all four attacking cases. The testing method is very straightforward and summarized as follows:

Step 1: Add the false data injections ΔD to corresponding load buses, then we have the new load vector D_{new}.
Step 2: Calculate the line flows using (10.72)

$$F_{\text{new}} = S \cdot (U \cdot G - V \cdot D_{\text{new}}) \tag{10.72}$$

Step 3: Compare the line flows after attacks with the original line flows before attacks. If the line flows in the non-attacking region do not change, the proposed method is verified.

It has been shown that the line flows in the non-attacking region do not change, which verifies all the cases discussed in this chapter. The results clearly show that the attacker can inject false data into load buses in the attacking region without impacting the line flows outside the attacking region. Thus, the injected false data meets $a = Hc$ for the entire network, which guarantees that the injected false data can pass the residual test of the control center. Thus, an attacker with limited attacking resources can choose one local region as the attacking target, obtain the network information (topology and parameter) of the chosen region, and compute the attacking vector to launch a successful local load redistribution attack.

10.6 Summary

Power grid security is crucial to a nation's economy and social welfare. However, the integration of information technology could make power systems vulnerable to cyber-attacks. In this chapter, a local attack model based on incomplete network information has been proposed. In our model, we show that the attackers only need to obtain the network information of the local attacking region to inject false data into smart meters in the local region of the power grid without being detected by the state estimator. Four attacking cases are analyzed in detail. Simulations on the modified IEEE 14-bus system confirm the effectiveness of the proposed model and algorithms.

Our works reveal that an attacker who has only parts of the network information (topology and parameters) of a power grid still can construct the valid attacking vector. This indicates that power systems are vulnerable to false data injection attacks, and thus, highlights the significance of studying attackers' possible attacking schemes and corresponding detecting and defending strategies.

References

1. Zhang Y, Wang L, Sun W (2013) Investigating the impact of cyber attacks on power system reliability. In: 2013 IEEE international conference on cyber technology in automation, control and intelligent systems, pp 462–467
2. Liang G, Weller R, Zhao J, Luo F, Dong Z (2017) The 2015 Ukraine Blackout: implications for false data injection attacks. IEEE Trans Power Syst 32(4):3317–3318
3. Guan Y, Ge X (2018) Distributed attack detection and secure estimation of networked cyber-physical systems against false data injection attacks and jamming attacks. IEEE Trans Signal Inf Process Netw 4(1):48–59
4. Liu Y, Ning P, Reiter MK (2009) False data injection attacks against state estimation in electric power grids. In: Proceeding of the 16th ACM conference on computer and communications security, pp 21–32
5. Deng R, Xiao G, Lu R, Liang H, Vasilakos A (2017) False data injection on state estimation in power systems—attacks, impacts, and defense: a survey. IEEE Trans Ind Inf 13(2):411–423
6. Abur Ali, Expósito A, Gómez (2004) Power system state estimation: theory and implementation. Univ Teknol Petronas 17(95):213–256
7. Yuan Y, Li Z, Ren K (2011) Modeling load redistribution attacks in power systems. IEEE Trans Smart Grid 2(2):382–390

8. Rahman MA, Mohsenian-Rad H (2013) False data injection attacks with incomplete information against smart power grids. In: Proceeding of the IEEE conference on global communications, pp 3153–3158
9. Liu X, Li Z (2014) Local load redistribution attacks in power systems with incomplete network information. IEEE Trans Smart Grid 5(4):1665–1676
10. Zimmerman Ray D, Carlos E, Murillo S (2010) MATPOWER: steady-state operations, planning, and analysis tools for power systems research and education. IEEE Trans Power Syst 26(1):12–19

Chapter 11
Optimal Attack Strategy on Power System

The topological information is usually kept confidential in the control center and not easy to obtain. For an attacker with limited attacking ability, it is difficult to have access to the admittance values of all the lines. Thus, the goal of an attacker is to determine an attacking region that requires the minimum amount of network information. However, it is a challenging issue for a practical size network. Our work in the literature [1] lays the theoretical foundation on cyber-attacks with only local information and provides clues to identify a small attacking region. In this chapter, we present a strategy to determine the optimal attacking region for one load bus built upon the local load redistribution (LR) attack theory in literature [1] by obtaining reduced network information.

11.1 Definitions of Terms

For the convenience of discussion, we first define several terms. Then, we study the topological characteristics of an attacking region, followed by detailed steps for determining the optimal attacking region [2].

Definition 11.1 (*LR-attackable and non-LR-attackable*) A measurement is LR-attackable if its reading can be changed with a certain range to satisfy KCL and KVL. A measurement is non-LR-attackable if its reading cannot be changed with a certain range to satisfy KVL. A network element (bus or line) is LR-attackable if its measurement can be LR-attackable. A network element is non-LR-attackable if its measurement is non-LR-attackable.

Note that, in LR attack, non-load buses (generator buses and zero-injection buses) are non-LR-attackable, and load buses and lines could be LR-attackable or non-LR-attackable. In literature [3], the study of cyber-attack object on overloading multiple lines is intensively carried out.

Y. Cao et al., *Cyber-Physical Energy and Power Systems*,
https://doi.org/10.1007/978-981-15-0062-6_11

Definition 11.2 (*effectively LR-attackable*) A network element is effectively LR-attackable if it is LR-attackable, and the attacking amount exceeds a given threshold value.

In literature [4], if the attacking amount at a bus is too large, the attack could be detected by the state estimator with a high probability. This is because the control center usually has pre-knowledge and experiences of the load distribution at buses. In this chapter, we assume that the maximum error of load forecast could reach 10–15% and, without loss of generality, the maximum allowable attacking amount (τ) of a load bus b is set to 15% of its load value. However, the attacking amount should not be too small either. Otherwise, the impact of the attacks on power systems operation could be insignificant. In this chapter, the threshold attacking amount of the load bus $b(\gamma_b)$ is set to 2.5–10% of its load value.

Definition 11.3 (*attacking region*) The attacking region of a load bus is defined as the region that satisfies the following conditions:

- The load bus is included in the attacking elements.
- If no line connected to a bus is attacked, then the bus is excluded from the attacking region. All lines connected to the bus are also excluded from the attacking region.
- If the two terminal buses of a line are included in the attacking region, then this line is also included in the attacking region.
- If a line is included in the attacking region, then its terminal buses are also included in the attacking region.

Definition 11.4 (*optimal attacking region*) The optimal attacking region of a load bus b is defined as a region that requires the minimum network parameter information, reduced topology and load level information, and the minimum number of measurements to be attacked to achieve an effective attack on bus b.

11.2 Modeling of Attacking Regions

11.2.1 Definition of LR Attacks

According to the definition of the attacking region, we can obtain three observations on the topological characteristics.

Observation 11.1 If bus i is a non-boundary bus in the attacking region, then all its neighboring buses and all lines connected to bus i must be included in the attacking region.

Proof We prove the observation by contradiction. Suppose that there exists at least one neighboring bus j not included in the attacking region. According to the definition of the attacking region, line $i - j$ is not in the attacking region, so bus i is either a boundary bus in the attacking region or a bus in the non-attacking region, which contradicts the fact that bus i is a non-boundary bus in the attacking region. Similarly, we assume that one line $i - j$ is not included in the attacking region, then at least one terminal bus of the line is not included in the attacking region. If bus i is not in the attacking region, then we obtain the contradiction. If bus j is in the non-attacking region, then bus i cannot be a non-boundary bus in the attacking region.

Observation 11.2 Suppose that bus i is a boundary bus in the attacking region. Line $i - j$ is included in the attacking region if bus j is a non-boundary bus or boundary bus in the attacking region.

Proof The proof is trivial. Since line $i - j$ is included in the attacking region, according to definition 11.3, buses i and j are also included in the attacking region as a non-boundary bus or boundary bus.

Observation 11.3 If bus i is in the non-attacking region, then all lines connected to bus i are excluded from the attacking region.

Proof The proof is trivial. Since bus i is in the non-attacking region, it is not attacked. So, bus i and all lines connected to bus i are excluded from the attacking region. Observation 11.3 has been proved.

We use Fig. 11.1 to illustrate Observations 11.1–11.3. The attacking region is circled by the solid rectangle in Fig. 11.1.

- **Observation 11.1**: Bus 1 is a non-boundary bus in the attacking region, so its neighboring buses 3, 4, 5 and lines 1–3, 1–4, 1–5 are all included in attacking region. Note that neighboring bus 5 is a boundary bus and non-attackable since it is a zero-injection bus, and neighboring bus 3 is a non-boundary bus and non-attackable since it is a zero-injection bus.
- **Observation 11.2**: Bus 4 is a boundary bus in the attacking region. Lines 4–1 and 4–5 are included in the attacking region since buses 1 and 5 are in the attacking region. Line 4–6 is not included in the attacking region since bus 6 is in the non-attacking region.

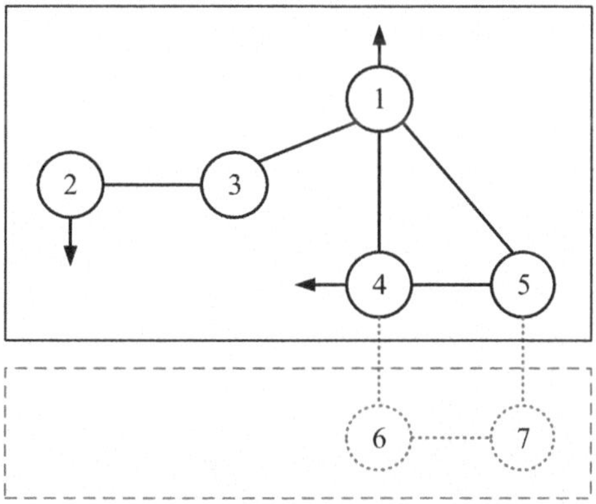

Fig. 11.1 Illustrative of topological characteristics of attacking regions

- **Observation 11.3**: Bus 6 is in the non-attacking region, so both lines connected to bus 6, i.e., lines 6–4 and 6–7, are excluded from the attacking region.

Observations 11.1–11.3 summarize the topological characteristics of an attacking region. These observations can be transformed into corresponding logic constraints such that a mixed integer linear programming (MILP) model can be built to represent the topological characteristics.

Observation 11.1 can be divided into two parts:

Part 1: If bus i is a non-boundary bus in the attacking region, then all its neighboring buses must be included in the attacking region.
Part 2: If bus i is a non-boundary bus in the attacking region, then all the lines connected to bus i must be included in the attacking region.

Part 1 is translated into the "if-then" condition: If $\delta_i = 1$, then $\delta_j = 1$ or $\upsilon_j = 1 \forall j \in \Omega_i$, which can be modeled as

$$\delta_i \le \delta_j + \upsilon_j \quad \forall j \in \Omega_i \tag{11.1}$$

$$\delta_j + \upsilon_j \le 2 - \delta_i \quad \forall j \in \Omega_i \tag{11.2}$$

Part 2 is translated into the "if-then" condition: If $\delta_i = 1$, then $w_{ij} = 1\ \forall j \in \Omega_i$ which can be modeled as

$$\sum_{j \in \Omega_i} w_{ij} \ge n_i \delta_i \tag{11.3}$$

Observation 11.2 is equivalent to the following "if-then" conditions:
If $\upsilon_i = 1$ and $\delta_j + \upsilon_j = 1$, then $w_{ij} = 1 \quad \forall j \in \Omega_i$
If $\upsilon_i = 1$ and $\delta_j + \upsilon_j = 0$, then $w_{ij} = 0 \quad \forall j \in \Omega_i$
which can be modeled as

$$w_{ij} \ge \upsilon_i + (\delta_j + \upsilon_j) - 1 \quad \forall j \in \Omega_i \tag{11.4}$$

$$w_{ij} \le (\delta_j + \upsilon_j) - \upsilon_i + 1 \quad \forall j \in \Omega_i \tag{11.5}$$

Since bus i cannot be both a non-boundary bus and a boundary bus in the attacking region, for all the buses in a power grid, we have

$$\delta_i + \upsilon_i \le 1 \tag{11.6}$$

It should be pointed out Constraint (11.12) can be discarded due to Constraint (11.6).

Observation 11.3 can be described using the "if-then" condition if $\delta_i + \upsilon_i = 0$, then $w_{ij} = 0 \quad \forall j \in \Omega_i$, which can be modeled as

$$\sum_{j \in \Omega_i} w_{ij} \leq n_i(\delta_i + \upsilon_i) \tag{11.7}$$

As discussed before, in the local load redistribution attacking model, an attacker does not need to know all the topology and parameters of the entire power grid. In this chapter, we design the attacking vector by making all the boundary buses in the attacking regions increase or decrease the same, that is, the non-attacking regions are treated as a connected network whether they are really connected or disconnected. In this case, the attacker does not need to know the topology information of the non-attacking region since it does not need to determine the connectivity of the non-attacking region(s).

The condition that all the boundary buses in the attacking region have the same incremental phase angle, e.g., α, can be modeled as

$$\Delta\theta_i + (M_1 + \alpha)(1 - \upsilon_i) \geq \alpha \quad \forall i \in \Omega_{BA} \tag{11.8}$$

$$\Delta\theta_i - (M_1 - \alpha)(1 - \upsilon_i) \leq \alpha \quad \forall i \in \Omega_{BA} \tag{11.9}$$

According to Theorem 9.1, all the buses in the non-attacking regions also have the same incremental phase angle α. This can be modeled as

$$\Delta\theta_i + (M_1 + \alpha)(\delta_i + \upsilon_i) \geq \alpha \quad \forall i \in \Omega_N \tag{11.10}$$

$$\Delta\theta_i - (M_1 - \alpha)(\delta_i + \upsilon_i) \leq \alpha \quad \forall i \in \Omega_N \tag{11.11}$$

11.3 Optimal Attacking Region

Based on the local attacking principle in literature [1], we adopt a heuristic algorithm to determine the optimal attacking region for a load bus b. The principle of the entire algorithm is described in Fig. 11.2. First, we determine an initial attacking region for bus b and set it as the searching region.

To ensure that the attacker only needs to obtain the reactance of the minimum number of lines, the number of lines in the searching region is minimized. Next, we obtain the true reactance of all the lines in the searching region. Then, we calculate the maximum attacking amount Q_b. If $Q_b \geq \gamma_b$, we begin to determine the optimal attacking region in the current searching region; otherwise, we need to expand the searching region until the condition $Q_b \geq \gamma_b$ is satisfied. The details of determining the initial attacking region, expanding the searching region, determining the maximum attacking amount, and determining the optimal attacking region are presented in the next sections. The reason that we gradually expand the searching region is to reduce the load and topology information required. In each expansion, an attacker

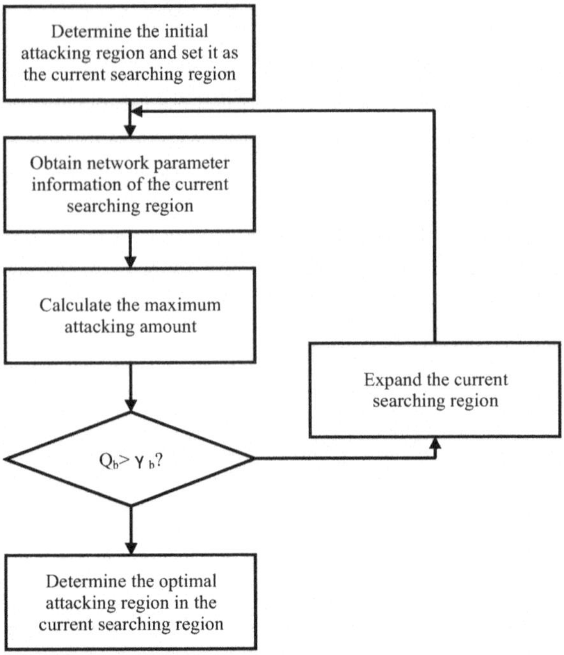

Fig. 11.2 Flowchart for determining the optimal attacking region

only needs to obtain the load and topology information of all boundary buses in the previous searching region.

The determination of an initial attacking region in this chapter is based on the following two observations.

Observation 11.4 It is highly likely that if bus b is LR-attackable in an attacking region with all the line reactances set to an arbitrary value, then bus b is still LR-attackable if the reactances of all the lines in the attacking region are set to their true values.

Observation 11.5 It is very likely that if bus b is LR-attackable in an attacking region, then bus b can be effectively LR-attackable in the region.

Observations 11.4 and 11.5 are made as a result of extensive experiments. In particular, Observation 11.4 is valid for all the experiments we have performed; Observation 11.5 is valid for over half of the experiments. According to Observations 11.4 and 11.5, if we determine an initial attacking region of bus b by setting the reactances of all lines to an arbitrary value, then there is a high probability that the initial attacking region would be a feasible attacking region. Thus, by doing so, for most load buses, we do not need to expand the initial attacking region to make it feasible. For the remaining load buses whose initial attacking regions are not feasible, the initial attacking regions are expanded until they are feasible.

Based on Observation 11.4 and without loss of generality, we can set the reactance of all the lines in the current searching region to one and calculate the susceptance matrix. Then, according to Theorem 10.1, we have

$$B_{S'}^{0}\Delta\theta_S = -V_S\Delta D_S \quad (11.12)$$

11.3.1 Algorithm of Determining a Feasible Attacking

The algorithm for determining the initial attacking region for a load bus $\boldsymbol{b}$ is summarized as follows:

Step 1: Obtain the topology and load level information of the primary attacking region. The searching region starts from the primary attacking region of the load bus.
Step 2: Set the reactance of all lines in the searching region to an arbitrary value.
Step 3: Set the incremental phase angles of all boundary buses, which include the buses in the searching region connected to the non-searching region, to be the same as α.
Step 4: Determine an attacking region by solving (11.13)

$$\min \sum_{l\in\Omega_{SL}} w_l \quad (11.13)$$

subject to

$$-\tau D_d \le \Delta D_d \le \tau D_d \quad \forall \mathrm{d} \in \Omega_{SD} \quad (11.14)$$

$$\Delta\theta_j = \alpha \quad \forall j \in\in \Omega_{BS} \quad (11.15)$$

$$\Delta D_d \ne 0 \quad (11.16)$$

$$B_{S'}^{0}\Delta\theta_S = -V_S\Delta D_S \quad (11.12)$$

$$(11.1{-}11.11)\forall i \in \Omega_{\mathrm{SB}}$$

The objective function in (11.13) is to minimize the number of lines whose parameters need to be known. Constraint (11.14) limits the attacking amount at buses. Constraint (11.15) ensures that all the boundary buses have the same incremental phase angle. Constraint (11.16) indicates that false data is injected into the measurement at bus b. Constraints (11.1)–(11.7) model the topological characteristics of the attacking region. Constraint (11.12) ensures that the injected false data obeys KCL and KVL. The optimization problem (11.13) is an MILP problem. If (11.13) is feasible, we find an initial attacking region, stop. Note that the initial attacking region may have less

number of buses and lines than the current searching region. Otherwise, if (11.13) is infeasible, go to step 5.
Step 5: Expand the current searching region.

We can see that the proposed algorithm is very simple yet very effective and has the following advantages:

- The determination of the initial attacking region does not require the true values of line reactance. The required network information is topology and load level information of the searching region.
- The initial attacking region is very efficient since it is also a feasible attacking region for most load buses.

11.3.2 Expansion Strategy

The searching region can be expanded by including all the neighboring buses of the boundary buses in the current searching region and the lines connecting the neighboring buses and boundary buses. In this process, the neighboring buses and the lines connecting the neighboring buses and boundary buses represent the topology information the attacker needs to obtain. In addition, the load level information of newly added buses is also obtained. Considering the limited capacity of an attacker, it is reasonable to assume that an attacker can obtain the parameters of at most K lines. Thus, during the expansion of the search region, we need to count the number of lines whose true parameters have been obtained. If it is greater than the given value K, the expansion will be stopped. Note that if the searching region cannot be further expanded and the attacking amount of load bus b has not been satisfied yet, then we can tell that load bus b is not effectively attackable by the attacker with limited attacking capacity.

The maximum attacking amount of load bus b is needed to determine whether the current searching region needs to be expanded. It can be obtained by solving the following linear programming (LP) problem given the true reactance of all lines in the current searching region.

$$Q_b = \max \Delta D_d \tag{11.17}$$

subject to constraint (11.12), (11.14)–(11.15).

11.3.3 Determine Attack Measurements

After the attacking region is determined, note that not all the measurements in the attacking region need to be attacks, thus we are supposed to minimize the number of measurement to be attacked. We can count the number of attacked measurements using the following models.

The optimal attacking region for a load bus b in the current searching region can be obtained by solving the MILP optimization problem (11.18).

$$\min \sum_{d \in \Omega_{SD}} \sigma_d + 2 \sum_{l \in \Omega_{SL}} \varphi_l \tag{11.18}$$

subject to

$$\Delta F_S = X_S^{-1} W_S^T \Delta\theta_S \tag{11.19}$$

$$B_{S'} \Delta\theta_S = -V_S \Delta D_S \tag{11.20}$$

$$\Delta D_b \geq \gamma_b \tag{11.21}$$

$$\Delta D_d + (\tau D_d)\sigma_d \geq 0 \tag{11.22}$$

$$\Delta D_d - (\tau D_d)\sigma_d \leq 0 \tag{11.23}$$

$$\Delta F_l + M_1 \varphi_l \geq 0 \tag{11.24}$$

$$\Delta F_l - M_1 \varphi_l \leq 0 \tag{11.25}$$

$$\Delta\theta_j = \alpha \quad \forall j \in\in \Omega_{BS} \tag{11.15}$$

The objective function in (11.18) is to minimize the number of attacking measurements. Note that there are two measurements for each line. Constraints (11.19) and (11.20) ensure that the injected false data obeys KCL and KVL. Constraint (11.15) represents the boundary condition. Constraints (11.22)–(11.25) count the number of attacking components.

11.4 Case Studies

We test the proposed optimal attacking region model using the IEEE 14-bus system. Bus 1 is changed from a non-attackable bus to a load bus. Loads at some buses are also modified for the purpose of illustrating the concepts in this chapter. Line reactance is the same as those in literature [1]. The system is composed of 14 buses and 20 transmission lines. The bus data can be found in the Appendix. We assume that this system is fully measured. That is we need one meter to measure the injection power for each bus and two meters to measure the power flow passing through each transmission line. Thus, 54 measurements are needed in total. The attacking magnitude for a load bus is limited at $\tau = \pm 15\%$ of the actual load. To ensure that the attacks can bring significant damages to power system, the attacking amount at a load bus must be greater than 10% of its load, that is, $\gamma_b = 0.1D_b$. Considering the limited capacity of an attacker, at most $K = 10$ lines are allowed in the searching region. We pick load bus 1 and load bus 12 to illustrate the detailed procedure.

11.4.1 Case 1: The Attacker Intends to Attack Load Bus 1

The searching region of bus 1 starts from its primary attacking region, which is composed of buses 1, 2, 5 and lines 1–2, 1–5, 2–5. Buses 2 and 5 are boundary buses between the searching region and the non-searching region. According to Theorem 9.1, to guarantee that there are no additional power flow exchanges between these two regions, the following boundary condition must hold:

$$\Delta\theta_2 = \Delta\theta_5 = \alpha \tag{11.26}$$

Obtain the loads of buses 1 and 5 and set the reactance of lines 1–2, 1–5, 2–5 to one, and then solve the optimization problem (11.13). Since (11.13) is infeasible and there are three lines in the current searching region, which is less than $K = 10$, we need to expand the current searching region.

$$\Delta\theta_4 = \Delta\theta_6 = \alpha \tag{11.27}$$

The new searching region consists of buses 1, 2, 3, 5 as non-boundary buses, buses 4, 6 as boundary buses, and lines 1–2, 1–5, 2–3, 2–4, 2–5, 3–4, 4–5, 5–6 whose reactance is set one. The loads of newly added buses 4 and 6 are obtained. According to Theorem 9.1, in order to make sure there are no power flows in or out of the searching region, the following boundary condition holds for buses 4 and 6, which are the boundary buses in the searching region.

Again, we solve (11.13), which is feasible now. The solution to (11.13) determines the initial attacking region, which is composed of buses 1, 2, 3 as non-boundaries, buses 4, 5 as boundary buses, and lines 1–2, 1–5, 2–3, 2–4, 2–5, 3–4, 4–5. Note that the initial attacking region is not the same as the new search region.

Then in order to calculate the maximum attacking amount of bus 1, we need to obtain the reactance of all lines in the initial attacking region, which is now the current searching region. The maximum attacking amount of bus 1 is determined by solving the following LP.

$$Q_1 = \max \Delta D_1 \tag{11.28}$$

subject to

$$\begin{bmatrix} 21.3840 & -16.9005 & 0 & 0 & -4.4835 \\ -16.9005 & 33.3743 & -5.0513 & -5.6715 & -5.7511 \\ 0 & -5.0513 & 10.8982 & -5.8469 & 0 \\ 0 & -5.6715 & -5.8469 & 41.8457 & -23.7473 \\ -4.4835 & -5.7511 & 0 & -23.7473 & 37.9499 \end{bmatrix} \begin{bmatrix} \Delta\theta_1 \\ \Delta\theta_2 \\ \Delta\theta_3 \\ \Delta\theta_4 \\ \Delta\theta_5 \end{bmatrix} = \begin{bmatrix} \Delta D_1 \\ 0 \\ \Delta D_3 \\ \Delta D_4 \\ \Delta D_5 \end{bmatrix} \tag{11.29}$$

$$-2.265 \leq \Delta D_1 \leq 2.265 \tag{11.30}$$

$$-3.33 \leq \Delta D_3 \leq 3.33 \tag{11.31}$$

$$-1.815 \leq \Delta D_4 \leq 1.815 \tag{11.32}$$

$$-1.14 \leq \Delta D_5 \leq 1.14 \tag{11.33}$$

$$\Delta\theta_4 = \Delta\theta_5 \tag{11.34}$$

Solving the above LP, we get the maximum attacking amount of bus 1

$$Q_1 = 2.2650\,\text{MW} > 1.5100\,\text{MW} \tag{11.35}$$

Thus, the current searching region is a feasible attacking region as it satisfies the attacking amount requirement. It can be seen that to determine the feasible searching region, an attacker needs to obtain the parameter information of seven lines marked in bold in Fig. 11.3, instead that of all 20 lines in the entire power grid.

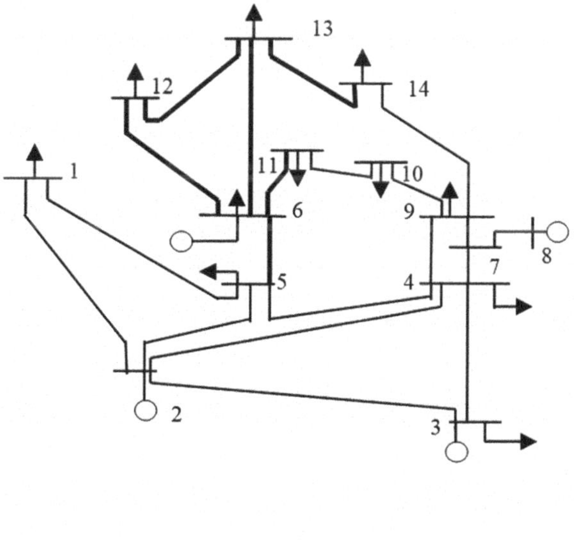

Fig. 11.3 IEEE 14 nodes network with load bus 12 attacked

Solving (11.18), we get the optimal attacking region which is the same as the feasible searching region, and the corresponding false injection data at load buses is found.

11.4.2 Case 2: The Attacker Intends to Attack Load Bus 12

The primary attacking region of bus 12 consists of buses 6, 12, 13 and lines 6–12, 6–13, 12–13. The following constraint holds for boundary buses 6 and 13 according to Theorem 9.1.

$$\Delta\theta_6 = \Delta\theta_{13} = \alpha \tag{11.36}$$

Obtain the loads of buses 6, 12, 13 and set the reactance of lines 6–12, 6–13, 12–13 to one and solve the optimization problem (11.13). Since (11.13) is feasible and there are three lines in the current attacking region, which is less than $K = 20$, we get the initial attacking region, which is the same as the current searching region. We then obtain the true reactance of lines 6–12, 6–13, 12–13 to calculate the maximum attacking amount of bus 12 by solving (11.18).

When $Q_{12} = 1.3679\,\text{MW} < 1.5000\,\text{MW}$, the initial attacking region does not satisfy the attacking amount requirement and needs to be expanded. Buses 6 and 13 are boundary buses in the initial attacking region, so we add neighboring buses 5, 11, 14 and lines 5–6, 6–11, 13–14 into the initial attacking region to form the new searching region. According to Theorem 9.1, the following boundary condition holds for buses 5, 11, 14, which are the boundary buses in the new searching region.

$$\Delta\theta_5 = \Delta\theta_{11} = \Delta\theta_{14} = \alpha \tag{11.37}$$

Obtain the true reactance of the newly added lines 5–6, 6–11, 13–14. Solving (11.13) again, we get the maximum attacking amount of bus 12.

$$Q_{12} = 1.6314\,\text{MW} > 1.5000\,\text{MW} \tag{11.38}$$

The attacking amount of bus 12 is increased and satisfies the condition of $Q_{12} > \gamma_{12}$. Thus, the expanded searching region is feasible. We can see that to determine the feasible searching region, an attacker needs to obtain the network parameter information of six lines marked in bold in Fig. 11.3, much less than the number of lines (20) in the entire power grid.

Solving (11.18), we get the optimal attacking region, which is composed of buses 6, 12, 13, 14 and lines 6–12, 6–13, 12–13, 13–14. Note that the optimal attacking region is only a subnetwork of the feasible searching region with bus 5 and lines 5–6, 6–11 being excluded.

Accordingly, the false injection data at load buses is

$$\Delta D_6 = -0.6000\,\text{MW};\ \ \Delta D_{12} = 1.5000\,\text{MW};$$
$$\Delta D_{13} = -0.9169\,\text{MW};\ \ \Delta D_{14} = 0.01696\,\text{MW};$$

Next, we calculate the post-attack power flows and compare it with the original power flows. If power flows in the non-attacking region do not change, the proposed method is verified. Table 11.1 lists the power flows before and after attacks. The third column represents the original line power flows before attacks, and the last two columns give the line power flows for case 1 and case 2, respectively. The changed power flows are marked in bold. It can be seen that the power flows in the non-attacking region do not change, and there are no additional power flows in or out of the attacking region. Thus, the local attacking scheme is successful and will not be detected by the traditional bad data test procedure.

Comparing cases 1 and 2, we find that the initial attacking region obtained using algorithm 1 could be feasible (case 1) or not feasible (case 2). Next, we want to make further investigation and find the percentage of initial attacking regions that are also feasible. The verifying method is straightforward. Suppose that there are *ND* load

Table 11.1 Power flow changes before and after attacks

Index	Line	Original flows (MW)	Case 1 (MW)	Case 2 (MW)
1	1–2	−22.6712	−23.8647	−22.6712
2	1–5	7.5712	7.2547	7.5712
3	2–3	3.2987	2.1053	3.2987
4	2–4	16.6034	16.6034	16.6034
5	2–5	17.4267	17.4267	17.4267
6	3–4	13.2987	14.6801	13.2987
7	4–5	2.4373	2.4373	2.4373
8	4–7	9.7629	9.7629	9.7629
9	4–9	5.6019	5.6019	5.6019
10	5–6	19.8352	19.8352	19.8352
11	6–11	−1.6965	−1.6965	−1.6965
12	6–12	9.0144	9.0144	9.6594
13	6–13	8.5173	8.5173	8.4723
14	7–8	0.0000	0.0000	0.0000
15	7–9	9.7629	9.7629	9.7629
16	9–10	7.6965	7.6965	7.6965
17	9–14	5.5683	5.5683	5.5683
18	10–11	5.1965	5.1965	5.1965
19	12–13	−5.9856	−5.9856	−6.8406
20	13–14	−4.5683	−4.5683	−4.5514

Table 11.2 Percentages of feasible initial attacking regions

System	Percentage			
	$\gamma_b = 0.1$ (%)	$\gamma_b = 0.05$ (%)	$\gamma_b = 0.025$ (%)	$\gamma_b > 0$ (%)
IEEE 24	70.59	82.35	94.12	100
IEEE 30	45.00	65.00	80.00	100
IEEE 39	47.62	61.90	71.43	100
IEEE 57	54.76	73.81	76.19	100
IEEE 118	60.61	78.79	88.89	100
Polish 2383	60.87	77.66	88.00	100

buses in a power grid and set $m = 0$. For each load bus b, the first determine the initial attacking region, then calculate the maximum attacking amount of bus b in the initial attacking region based on true line reactances and increase counter m by one if $Q_b \geq \gamma_b$. Then, the percentage of feasible initial attacking regions is given by

$$p = \frac{m}{ND} \times 100\% \tag{11.39}$$

Table 11.2 gives the percentages of feasible initial attacking regions for IEEE 24-bus system, IEEE 30-bus system, IEEE 39-bus system, IEEE 57-bus system, IEEE 118-bus system, and the Polish 2383-bus system [5].

It can be observed that the initial attacking regions of about 45% or more load buses have an attacking amount exceeding the required amount when $\gamma_b = 0.1$. As γ_b decreases, more initial attacking regions of load buses are feasible, which indicate that for most load buses, an attacker does not need to expand the initial attacking regions by obtaining the true line reactance of extra lines. Thus, the number of lines whose reactance the attacker must know is minimized. Moreover, if the strict requirement on the attacking amount of load buses is relaxed, i.e., $\gamma_b > 0$, all the initial attacking regions determined by algorithm 1 are also feasible attacking regions. So, Observations 11.4 and 11.5 are verified.

In this section, we investigate the number of lines whose parameter information is needed to determine the optimal attacking region based on the proposed strategy, i.e., the minimum number of lines to launch a successful local false data injection attack without being detected. The average numbers n for different systems are given in Table 11.3. The value in the bracket denotes n as a percentage of the total number of lines. It can be seen that n is much smaller than the total number of lines in a system. Moreover, as the size of a system increases, n remains almost unchanged and accordingly the percentage would decrease. For instance, the value of n falls between 6 and 11 when $\gamma_b = 0.1$, between 4 and 8 when $\gamma_b = 0.05$. When $\gamma_b > 0$, an attacker only needs to obtain the parameters of three lines on the average to attack a load bus. An attacker needs to obtain the parameters of average 20.2% lines to attack a load bus for IEEE 24-bus system when $\gamma_b = 0.1$. However, for the large-scale Polish system, only the parameters of 0.34% lines are needed. In addition, n

Table 11.3 Average number of lines in the optimal attacking region

System	Percentage			
	$\gamma_b = 0.1$ (%)	$\gamma_b = 0.05$ (%)	$\gamma_b = 0.025$ (%)	$\gamma_b > 0$ (%)
IEEE 24	20.2	16.3	11.0	11.0
IEEE 30	25.9	18.8	14.2	8.8
IEEE 39	18.5	15.6	12.4	8.4
IEEE 57	10.5	7.2	6.6	4.1
IEEE 118	4.9	3.4	2.3	1.6
Polish 2383	0.34	0.23	0.18	0.09

decreases when the given threshold value γ_b becomes smaller. This is because the attacking amount requirement of a load bus can be satisfied more easily when γ_b gets smaller.

11.5 Summary

Cyber security has emerged as a critical issue in smart grid development. In this chapter, we investigated the topological characteristics of an attacking region and built the corresponding MILP model. Based on the understanding of the topological characteristics of the attacking region, we took the first attempt to propose a strategy to determine the optimal attacking region of a single load bus by obtaining less network information.

As the extension of the current work, we will next explore the strategies of determining the optimal attacking region(s) for multiple load buses. The simplest approach is to determine the optimal attacking region for each load bus independently and then combine the optimal attacking regions of all load buses as the final attacking region. However, this may lead to the unnecessary efforts of obtaining the parameters of more lines. Thus, the effective methods need to be explored in the future. Possible directions are to develop better heuristic rules, consider the connectivity of attacking regions, or adopt distributed attacking schemes and so on.

References

1. Liu X, Li Z (2014) Local load redistribution attacks in power systems with incomplete network information. IEEE Trans Smart Grid 5(4):1665–1676
2. Liu X, Bao Z, Lu D, Li Z (2015) Modeling of local false data injection attacks with reduced network information. IEEE Trans Smart Grid 6(4):1686–1696
3. Tan Y, Li Y, Cao Y, Shahidehpour M (2018) Cyber-attack on overloading multiple lines: a bilevel mixed-integer linear programming model. IEEE Trans Smart Grid 9(2):1534–1536

4. Yuan Y, Li Z, Ren K (2011) Modeling load redistribution attacks in power systems. IEEE Trans Smart Grid 2(2):382–390
5. Zimmerman R, Murillo-Sanchez C, Thomas R (2011) MATPOWER: steady-state operations, planning, and analysis tools for power systems research and education. IEEE Trans Power Syst 26(1):12–19

The manufacturer's authorised representative in the EU is Springer Nature Customer Service Centre GmbH, Europaplatz 3, 69115 Heidelberg, Germany. If you have any concerns regarding our products, please contact ProductSafety@springernature.com

Printed and bound by CPI Group (UK) Ltd, Croydon, CR0 4YY
30/09/2026
02225452-0006